CONCEPTS IN
WINE
TECHNOLOGY

YAIR MARGALIT

Concepts in Wine Technology

Yair Margalit

The Wine Appreciation Guild

San Francisco

Concepts in Wine Technology

A Wine Appreciation Guild Book

Text copyright © 2004 by Yair Margalit, Ph.D.

Editorial Staff: Jeff Brown, Bryan Imelli, Alex Shaw
Text Design: Diane Hume
Cover Design: Alison Wood, Woodland Graphics

Library of Congress Cataloging-in-Publication Data
Text by Yair Margalit, Ph.D.
Concepts in Wine Technology
1st ed.,
ISBN 1-891267-51-5
1.   Wine and winemaking—Science—Chemistry I. Yair Margalit, Ph.D.
Library of Congress Control Number: 2003113072

Manufactured in the United States of America

The Wine Appreciation Guild, Ltd
360 Swift Avenue
South San Francisco CA 94080
(650) 866 3513
www.wineappreciation.com

Contents

FOREWORD

After the book "Concepts in Wine Chemistry" was published in 1997, I felt that another book, more technically oriented, would better cover the practical aspects of wine making. The first book gives the theoretical background of wine, and this one, brings the technology of practicing it. In fact, this book replaces the old one, "Winery Technology and Operation" which was first published in 1990. Instead of printing a new edition, we decided to change it to a different approach to the subject, because many things have changed since its first publication. As a wine technology book, it stands by itself.

But we felt that some parts in the chemistry book have changed as well, and are important enough to be included in this technical book. If the reader is interested in more details and in digging deeper into certain matters, please refer to the chemistry book.

This book contains seven chapters starting with grape ripening and pre-harvest operations, through harvest, fermentation, cellar operations, barrel aging, bottling, and a chapter on general aspects of winemaking, which includes sulfur-dioxide, wine spoilage, legal aspects, and phenolic compounds. Also, we've included an appendix, which contains our view on wine evaluation, a list of other higly recommended books related to wine and winemaking, and a subject index.

—Yair Margalit

Chapter I

Pre-Harvest

CHAPTER I: PRE-HARVEST

A. Grape Ripening

The main object in following up the developments and changes of the grapes during the ripening period is to confirm the maturity state of the grapes, with respect to the wine we are going to produce. The maturity definition of the grapes (or more specifically the time to harvest) is not a simple and clean-cut decision; it depends on many factors, such as the variety of the grapes, the appellation (soil and climate), the weather during the last days or weeks before the prospective time of harvest, the bunch health (mold or insect infection), the vineyard disease state, and the style of wine to be made.

The development of the berries from the time of fruit set until ripeness is roughly sketched in the following figure. At the beginning of stage III (Veraison), the berries begin to change color and soften. Just after these changes, the development of the berries is very fast, followed by rapid increase in the sugar concentration. Stage IV, where the grapes are over-ripe, is characterized by further softening of the berry's skin, shriveling due to water evaporation, and finally drying or mold infection. The exact time to harvest, as will be discussed in this chapter, is somewhere in the interval between the end of stage III, and up to a certain point in stage IV.

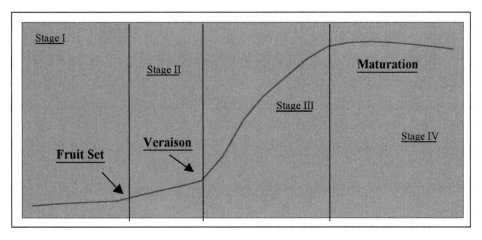

The figure above shows the four stages of grape development. The berry parameter (y-axis) might be the berries' weight, or sugar concentration. The major changes occur during stage III, when frequent samples of the maturing grapes should be taken and measured.

The chemical changes one should follow are: sugar concentration, acid content, and pH level. Start following these changes when the grapes' color begins to change. White grapes will turn from a deep green to yellow-green, and red grapes from green to purple-red. The change is quite rapid, a matter of a week or two. These changes usually happen at a sugar level of about 15–17 Brix.

1. Sugar

The main sugars in grapes are D-glucose and L-fructose, generally in 1:1 ratio, with fluctuations of ±30%, depending on the variety and maturity stage of the grapes. In grapes infected with *Botrytis Cinerea* mold, the ratio is in favor of fructose, which is twice as sweet as glucose. During fermentation with most yeast strains, the consumption of glucose is faster than that of fructose, and toward the end of fermentation, most of the residual sugar is therefore fructose. Other sugars which are found in grape juice are sucrose (about 2–10 gr/L), L-rhamnose (0.2–0.4 gr/L), L-arabinose (0.5–1.5 gr/L), and pectin which is a higher molecular weight sugar at about 0.2–4.0 gr/L. Actual range of glucose and fructose concentration in grape juice is 80 – 130 gr/L each.

Measuring Sugar Content in Grape Juice

Being the major soluble solid content in grape juice (above 95%), the measure of must density will reflect fairly well the total concentration of sugars in the must. From the tradition of the trade we have inherited three sugar concentration units, namely, *Baumé, Brix* and *Oechsle*.

Baumé is approximately the potential alcohol (in ml/100 ml wine), which will be produced by fermenting the sugars to dryness. So, e.g. a must with 12 Baumé will have about 12% alcohol. Originally this unit was designed to measure solid concentrations in solutions (weight/weight) and was calibrated by sodium chloride. Its use in the wine industry reflects the potential alcohol content in %(vol/vol).

Oechsle is a direct hydrometer reading translated to a more workable number, namely, the density of the must minus one, multiplied by 1000. Or in a short notation **Oechsle = (d - 1.000)x1000** (where: **d** is the density). For example if the density of a must sample is 1.074 (which corresponds to 9.9 Baumé), its Oechsle value is (1.074 - 1.000)x1000 = 74.

Brix (B°) (also called Balling) is the % of the total solids in solution, in grams of solute/100 gram solution (gram/gram). This unit is very much used in general chemistry practice but in recent years it is the most used unit for sugar content in the wine industry. It will also be used in this book. Baumé is still in use mainly in France, and Oechsle in Germany and Switzerland. The relations between the density (at 20°C), vs. the Baumé and Brix are shown in the following figure:

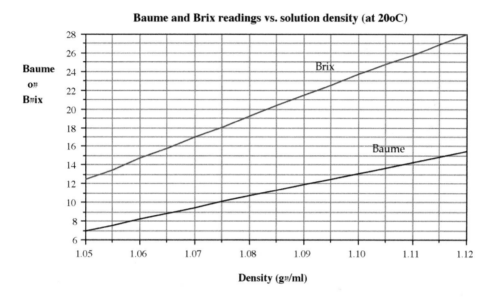

In order to use the above figure for Brix-Baumé transformation, one should take the actual number in the measured system (Brix or Baumé) in the graph and follow in the vertical direction (y-axis) to the other system to read its equivalent value. For example, 10 Baumé leads upwards to its equivalent value 18 Brix, or 26 Brix leads downwards to 14.5 Baumé.

The special hydrometers, which measure *directly* the Brix reading (or Baumé) are calibrated at 20°C. For measuring must at a different temperature, a correction should be made. At temperatures above 20°C the correction value should be added, and below 20°C the correction value should be subtracted. The following correction figure might be helpful:

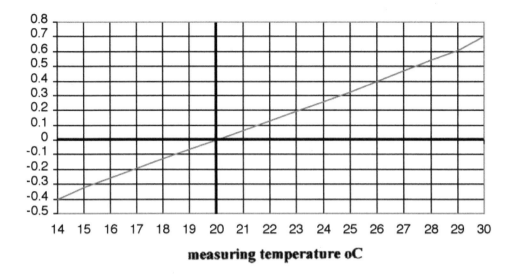

measuring temperature oC

For example if you read 22.3B° at 26°C, then the correct Brix content is 22.7B°. Since the Brix (measured by density) includes all solid materials in the must (sugar, acids, salts, proteins, pigments, etc.), the approximate percentage of pure sugar content (within ± 0.5% accuracy) can be estimated. At the relevant Brix (B°) region in stages III and IV (18B°–25B°) the pure sugar concentration may be given by:

$$\textbf{\% Sugar (w/v) = (Brix – 2.1) x density} \qquad \text{(1)}$$

Note that the % sugar is expressed by weight per volume.

The sugar is converted by the fermentation, mostly to ethyl alcohol and carbon dioxide:

$$C_6H_{12}O_6 \longrightarrow 2\,C_2H_5OH + 2\,CO_2$$
$$\text{180 gram} \qquad\qquad\qquad \text{92 gram} \qquad \text{88 gram}$$

From the molar ratio of the products to the substrate (sugar), 51% of the sugar is transformed into ethyl alcohol. In practice, the yeast itself utilizes part of the sugar, and some part is converted to other products (higher alcohols, aldehydes, esters, acids, etc.). Therefore, the molar yield of ethyl alcohol production during fermentation is around 47%. In order to have a good estimation of the potential alcohol content in the finished wine (expressed in the common volume-per-volume unit), one can use the relation:

> **Each 1.75% sugar (w/v) --> 1% alcohol (v/v)** (2)

By combining expressions (1) and (2), an expression connecting the potential alcohol to the Brix-density relation can be derived:

> **% alcohol (v/v) = 0.57 x (Brix – 2.1) x density** (3)

This expression is valid in terms of expression (1) validity, namely, in the Brix range of $18B°-25B°$, within accuracy of ±0.2% of the potential alcohol. Thus, for example a must with $22.7B°$ and density of 1.095 has a potential alcohol of 12.8±0.2%. The final ethanol content is also dependent on the fermentation temperature due to evaporation losses.

Another tool for measuring the sugar concentration is the refractometer, which is based on the refractive index of sugar (linear dependence on concentration). This tool is handier than the hydrometer (which measures the must density), and therefore very useful for field tests. It's values are slightly lower than those measured by the hydrometer.

2. Acids

We shall start with some words about acidity, to emphasize its importance in grapes and wines. Acidity is one of the most influential factors in wine. It affects the wine's microbial stability, malo-lactic fermentation, color and aging rate, as well as its tartarate and protein stability. But above all, acidity has a major effect on wine's tasting balance perception.

Acidity in wine has two sources: acids that develop in the grapes (mainly tartaric, malic and citric), and acids formed through the process of winemaking (lactic, succinic, acetic, and others). We shall mention here some of the major acids in grapes and wine, but will start with the general aspects of acidity relevant to winemaking. Grape acids are considered to be weak organic acids, and the following section will deal with the behavior of such acids in solution.

The ionic dissociation of week acid represented as **AH** is:

$$AH \iff A^- + H^+$$

The *dissociation constant* K_d is defined as: $K_d = [A^-][H^+]/[AH]$; where $[A^-]$ and $[H^+]$ are the equilibrium concentrations (molar concentration) in solution of the anionic form of the acid, and its proton respectively. $[AH]$ is the undissociated acid concentration. The K_d expresses the strength of the week acid, namely, greater K_d (which mean the concentrations of $[A^-]$ and $[H^+]$ are higher), means stronger acid, if compared with another one, which has a smaller K_d (lower concentrations of $[A^-]$ and $[H^+]$). In other words, the strength of an acid is a measure of its ability to release proton ions $[H^+]$ into solution.

The logarithm of K_d with minus sign is defined as pK_a, namely:

$$- \log K_d = pK_a$$

The pK_a expresses the acid strength in a different way, namely, the lower the pk_a, the stronger the acid (because of the minus sign of the logarithm).

The following expression is a very useful one, connecting the intrinsic potential of a week acid to dissociate, and expressing its acidity at any given pH:

$$\boxed{\log [A^-] / [AH] = pH - pK_a}$$

The logarithm of the ratio of the anionic part of the acid $[A^-]$ to the un-dissociated one $[AH]$, is equal to $pH - pK_a$. One can use this expression to calculate the ratio $[A^-]/[AH]$ at a given pH, or to calculate the pH by knowing that ratio of each acid. In wine, the pH is the most important factor, as it reflects the actual proton concentration in solution, which is not necessarily the total acid concentration. In strong inorganic acids, where dissociation is practically 100%, the pH (in log scale) is equal to their concentration. But in weak acids, where dissociation is partial, the above equation defines the connection between the actual acid concentration, and its pH. Most wine acids are di-acids (capable of releasing two protons per one molecule), so a second dissociation takes place with a second dissociation constant, which is always very small compared to the first one:

$$K'_d \qquad\qquad K''_d$$
$$AH_2 \Longleftrightarrow H^+ + AH^- \Longleftrightarrow H^+ + A^= \quad ; \ or \ \{AH_2 \Longleftrightarrow A^= + 2H^+\}$$

In grape juice and wine there are many organic week acids, which together contribute to the wine acidity and its pH. Each one of the acids has its typical pK_a and its actual concentration. Together with another major factor, namely, the concentration of the alkaline metal potassium, the wine pH is established. The calculation of the expected pH in such medium containing week acids and alkaline metal is beyond this chapter but it is helpful to know the final result, that the combinations of the total acids and potassium cation in wine, bring its pH range to span between 2.8 up to 4.0 .

White wines usually have a lower pH (3.0–3.5), while red wines usually have higher pH values (3.3–3.9). Sparkling wines generally fall in the very low pH range (2.9–3.1) because grapes for sparkling wines are harvested early, before they reach full maturity, when their pH is still very low.

It is important to distinguish between the terms *total acidity* and *titratable acidity*. The total acidity is the content of the *total organic acids* concentration in wine. The titratable acidity is the total *proton concentration* in wine, which is measured by titration with a strong base. It is not equal to the total acidity. It is less. The difference accounts for the alkaline potassium, which partially neutralizes the organic acids. The total acidity is therefore equal to the sum of the concentrations of titratable protons and potassium. Also, there is some contribution from sodium, which is contained in must, in lower concentrations than potassium. Therefore:

The total acidity (molar) = $[H^+]_t + [K^+] + [Na^+]$; where $[H^+]_t$ is the titratable acidity, and $[K^+]$ and $[Na^+]$ are the alkaline metals concentrations (molar).

The combination of weak organic acids and alkaline metals in wine makes it a buffer to pH changes. The buffering capacity depends on the acids and alkaline-metal ions concentrations.

Grape Acids

Tartaric acid is the key acid in wine. It is quite unique in nature, being found solely in vine fruits. Its formal presentation is COOH – CH(OH) – CH(OH) – COOH, which hides its uniqueness as being symmetrical on one hand, and containing two unsymmetrical carbon atoms in its structure on the other hand. The last character causes this molecule to exist in four stereoisomers. Two of them are mirror images of each other and are defined as optical isomers. The great wine chemist Pasteur was the first to found this unique and very important chemical phenomenon in the mid-nineteenth century. Only one of these optical forms (out of the four possible isomers) is found in grape juice. Its notation is (L+) tartaric acid.

Malic acid is also a major acid in grape juice but it is a very common acid in many fruits. Its formal presentation is COOH – CH(OH) – CH_2 – COOH which has only one asymmetric carbon and hence only two isomers. The natural grape malic acid is the (L-) malic isomer. Before ripening the two acids are at about the same concentration. During the advanced maturing process, toward the end of stage III, the concentration of malic acid gradually and consistently decreases by respiration, while that of tartaric acid remains practically unchanged. At full maturity the concentration of tartaric acid is always higher, where the tartrate/malate ratio is found mostly in the range of 1.1–2.0 depending on the variety, climate, and maturity state. As a result of the malic acid respiration during maturation, the total acidity is reduced, and at the same time also the concentration of potassium is increased. These two processes cause the grape juice pH to increase, mainly at the end of the ripening process.

L-citric acid is another acid which exists in must in a comparable amount, with a concentration range between 0.1–0.3 gr/L.

The total titratable acidity in must is in the range of 4 to 12 gr/L. The dissociation constants (at 25°C) and the corresponding pK_a's for the three major acids in must are presented in the following table:

Acid	Dissociation Constant	pK$_a$
tartaric	11.0 x 10^{-4}	2.96
malic	4.0 x 10^{-4}	3.40
citric	8.4 x 10^{-4}	3.07

It is clear that tartaric acid, which has the lower pK$_a$ is the strongest one, and being also the major acid in must it is the most influencing on the must and wine acidity. Notice that the two major acids, tartaric and malic, are di-acidic, which mean that the second proton also contributes to their total acidity but not to the pH because the second dissociation constant is by far very low (4.25x10^{-5} and 7.9x10^{-6} respectively).

3. Maturity

There are some different criteria of how to determine the maturity stage of the grapes and therefore the optimal time for harvesting. It depends very much on the grape variety, the type of wine being produced, and the climate and microclimate of the vineyard. It is also very dependent on the annual climate at each specific vintage. Harvest time therefore might change drastically every year at the same vineyard. In any method being used to determine maturity, there is practically no need to start to collect grape samples before the grapes have reached 16°–17° Brix. The profiles of the major parameters during the weeks of final maturation have the following shapes:

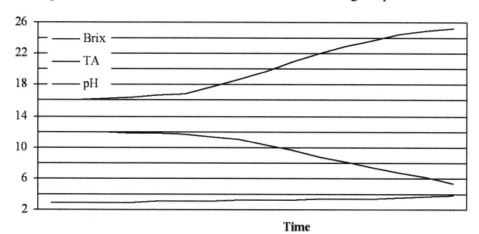

The Brix, TA and the pH profiles during the final states of maturation (stages III and IV). The Y-axis represents the actual range values of these parameters.

By measuring the changes in the sugar, TA and pH over certain time intervals, one can follow the grapes' development towards maturity. Yet the main question is still unanswered: when to harvest?

If the sugar is allowed to develop unchecked, other parameters could be damaged. For example, the TA might be too low, the pH too high, and/or the aroma materials might change their nose to a jammy or baked flavor. Also, waiting too long to harvest might expose the grapes to weather risks in certain areas, or insect and mold infections. There must be a certain point, which is the optimal one, when the considerations are best taken into account. When is that point?

We shall skip all other methods found in past literature and bring here the one that is, in our experience, the most valuable and practical. It considers mainly the sugar development. As long as sugar is still accumulating (which of course is not always the case), all other parameters are more or less moving in the right direction. At each sampling, calculate the Brix /day accumulation. When it comes close to 0.1 Brix/day, start to think about harvesting. If it reaches 0.05 Brix/day, the maturation is close to being stuck and no further waiting is advised. This method is simple and, in most cases, satisfactory.

In cold growing regions where grape ripening is a well-known problem (e.g. Germany, certain places in France, and other northern countries), the major wine quality parameter is the Brix content of the grapes at harvest. Also in other places during certain years, weather conditions (cool summer, cloudy skies) may sometimes cause a delay in the ripening. If the delay is accompanied by rain and high humidity, the grapes will probably be infected with mold, which causes more problems as the harvest is delayed. Under such conditions there is no alternative but to harvest immediately, before the sugar and flavor have fully developed. The deficiency in sugar level and high acidity may be adjusted, but the aroma and flavor of the wine may suffer. Unfortunately, such wine is likely to be of low-to-medium quality. In these cases the sugar content should be watched very carefully in hope that it will reach the desired concentration. Even in medium and warm countries, it may happen that the maturity of the grapes is halted for unknown reasons. The Brix does not go up, the TA is continuously reduced, and the pH begins to run sky high. Waiting in such situations, with the hope of developing more sugar, is probably a mistake. The quality of the grapes will probably just decline.

In view of the above-mentioned hazards, it is very important to watch the smooth development of the Brix until it shows signs of slowing. This is vital for finding the right time to harvest.

There are some general guidelines to help us to relate the different parameters according to grape variety. These parameters depend very much on the appellation where these varieties are grown, and should be studied in each case. As a very general guide, here is the recommended *range* for the Brix, TA and the pH for white and red grapes:

Type of wine	Brix	Total Acidity (TA)	pH
White	19–23	0.7–0.9	3.1–3.4
Red	21–24	0.6–0.8	3.3–3.8

If in white grapes, for example, the Brix is still low (around 18 Brix), the TA has dropped down to below 0.6, and the pH is above 3.4, one should probably wait for the Brix to rise, and make the necessary corrections for the deficiency in acid and high pH. On the other hand, with the same TA and pH, if the Brix were close to 21–22, it would be better to harvest very soon.

Another, most important aspect to consider when determining grape maturity is the varietal aroma and flavor. With this in mind, the Brix level need only be kept within the general recommended range. Brix limits are important only for maintaining the alcohol range in the finished wine, typical to that variety or style of wine. The full varietal flavor of the grape is most important for achieving the maximum varietal character in the finished wine. While this sounds great, this method to achieve it requires experience and the ability to project the "grapes' flavor" and taste into the finished wine. Speaking from personal experience, this quality projection cannot be generalized and written in a book.

There are some more complicated methods which reflect the ripening development, such as measuring the tartaric/malic acids ratio, the polyphenolic compounds development, or the potassium development, but none of these methods include a conclusive, precise assessment of the maturity timing.

4. Sampling

The technique of sampling is very crucial in order to get a real representation of the grape's degree of ripeness. First, the distribution of the major components in the berries along the cluster is not even. The top berries in the cluster are higher in sugar than the bottom ones. After veraison it can be seen a distribution of color differences among the berries. In many cases even at full red color, some berries are still green. The clusters themselves are not even in their development, depending on their location on the vine, the amount of light they are exposed to, and the location of the vine in the vineyard. On top of this, the content of the berry is not uniformly distributed. Usually the sugar concentration is higher closer to the skin and drops down toward the center of the berry. The acid's concentration follows the opposite trend. All the above factors are important to realize, in order to get the best representatives of must sample from the vineyard.

The sampling technique which might give quite a good picture on the grapes maturity is to collect about 50 - 100 clusters, from different locations on the vine (both sides of the row), lighted and shadowed, and from different vines in the area. It is advisable not to collect the sample from a very large area of the vineyard (not more than 5 acres per sample), because of the wide variability due to microclimate and soil in the same vineyard. The simplest way to pick the clusters is to walk in the vineyard along the rows and every certain number of vines to pick a cluster from both sides of the row. The sample should be weighed and the number of clusters be counted in order to establish the average weight of a cluster. Counting the average number of clusters on each vine, may lead to an estimation of the yield in that vintage. The sample has to be crushed by hand as firmly as possible, and the juice separated from the skins and seeds by a net. The Brix, total acidity and pH should be measured and be placed on a graph. The Brix/day change since the last sampling should be noticed. The sampling and measurements should be taken once or twice a week, depending on how close the grapes are to finishing their maturation process.

5. Composition and Yield

The overall weight composition of the grape cluster is:

	% (Range)	Average %
stems	2–6	4
seeds	0–5	4
skins	20–30	25
must	60–70	67

The lower figure in the must range (60%) is for the free-run and very light press juice, and the upper one (70%) includes the fully pressed juice.

As for the must, its composition is:

	% (Range)	Average %
water	75–85	77
sugars	17–25	22
organic acids	0.4–1.0	0.6
other dry extract	0.3–1.0	0.5

The other dry extract contains proteins, amino acids, esters, alcohols, polyphenols, minerals, and aroma components.

From the wine production aspect the yields are:

White Wine: about 60% of the grapes' weight (made out of free-run only) and close to 70% of the grapes weight (when press-run included), ends up as white wine.

Red Wine: about 60% of the grapes' weight (free-run) and up to 65% (press-run included), ends up as red wine.

These figures of production yield can be expressed in a different way: from every ton of grapes (1000 Kg) red or white, one should get about 60 to 65 cases of free-run wine bottled in 750 ml. bottles, at 12 bottles per case. If press-run is included, another 3–7 more cases can be expected.

B. Pre-Harvest Operations

The winery's policy for the coming vintage should be set up before planning the major aspects of winery operation. This should include the following points:

- Types and qualities of wines to be produced.
- Quantities of each type.
- Grape sources (quantities, quality and prices if bought from other growers).
- Financial and budget plans.
- Marketing prospects and strategies.
- Availability of manpower for the crush season.

1. Vineyard Management

From veraison time up through harvest, vineyard management has a close relation with wine quality and production. For best results, the winemaker should pay close attention to:

Chemicals used against different kinds of insects and fungus: Some chemicals may have residual activity causing health hazards, and some (especially those preventing fungus infections) may inhibit fermentation. When sulfur dust is used, (against powdery mildew and other fungus agents) and if is carried by the grapes into the wine, it may cause production of hydrogen sulfide, which has an unpleasant odor. The chemicals are mainly used as preventive agents during shoot and leaf growth (March-July in the northern hemisphere) and are needed much less toward the end of the summer. At this time the main concern is the bunch rot caused by different insects, and microbial agents such as *Botrytis Cinerea, Aspergillus, Penicillium, Acetobacter* and others. "Bordeaux Soup" (copper sulfate, widely used as a repellent against downy mildew in Europe), will carry copper into the must, and may later cause copper haze in white wine. (If the concentration of copper is below 0.5 ml/L, the haze will not be formed, but at the same time it may precipitate the H_2S in a "stinking" fermentation. More details on copper effects will be found later in this book.) More efficient chemicals are used now, which can focus on their specific targets.

In order to prevent all these chemicals from being carried onto the grapes and into the wine, their use should be terminated sometime after veraison (color changes of the grapes). The time remaining before harvest should be enough for the chemical agents to either decompose, evaporate, or be swept away by wind or rain.

Irrigation during the summertime when there is no rain: If irrigation is continued up to harvest, it may dilute the dry extract of the berries and reduce grape quality. It is believed (with no firm evidence), that the vine should struggle and be under some stress during the last period of ripening in order to produce a good quality wine. It may be true or not, but in any case, extra water at this stage will be used by the vine mostly for shoot growth, which does not contribute to the berry's development. Therefore, it is generally recommended to stop irrigation at least a month before the harvest. Care should be taken in hot summers or in hot regions, because drying the vines too early may cause the leaves to turn yellow and to stop photosynthesis. Consequently, the maturity could be delayed or even worse, not come at all. Therefore, very careful inspection should be carried out with regard to irrigation management. Probably the best method is to irrigate small quantities once or twice a week, just enough to keep the vine from turning yellow. Also, it might be a good idea to cause the vine to dig its roots way down into the ground, reaching for last winter's rainfall water instead of irrigation water. This can be done over a few years (better when the vines are still young) by managing long intervals between each irrigation, while using large quantities of water each time.

Leaf removal and pruning: After veraison, it's important that the clusters are not over-shadowed. Berries exposed to light develop better flavor and higher sugar levels than shaded ones. One efficient way to improve light exposure is to remove the leaves over grape clusters. It's a labor-intensive operation, but it leads to higher grape quality. Leaf removal should start after veraison, and should be repeated whenever necessary.

Another factor to consider is pruning. Long shoots on the vine do not contribute to the berries' development. They also create obstacles for hand harvesting as well as mechanical harvesting. Thus, it is good practice to prune long shoots, clearing the way in the vineyard and decreasing shadow areas on the vine. It is recommended that this pruning be done about 2 to 4 weeks before harvest, depending on the vigor growth of the vines.

Harvest: The traditional harvest is done by hand and, for small wineries, is much preferred over mechanical harvesting. Mechanical harvesting is used mainly by large wineries because of the low availability of manpower for hand harvesting, which is hard seasonal work with low pay. In addition, mechanical harvesting allows the grapes to be picked at night when temperatures are low, which is a big advantage in warm regions. However, the main disadvantage of mechanical harvesting is that most of the berries' skins are broken, releasing juice into the container. This can lead to oxidation and unwanted fermentation. Yet the most threatening factor of a mechanical harvest (besides temperature) is the time allowed to pass between the harvest itself, and the processing of the grapes in the winery. The shorter the time, the lower the risk of damage.

On the other hand, with hand harvesting, damage to the berries is minimal, so oxidation and uncontrolled fermentation are almost entirely prevented. Yet some precautions should still be taken. For example, when hand harvesting, it is recommended to start picking very early in the morning and end work by noon, so the grapes can be picked at cooler temperatures.

2. Preparing the Winery for Harvest

The mid summertime before the harvest is the best time to prepare the winery for its new vintage. Everything needed for the harvest should be made ready. This is vital because during harvest there is no time to look for missing chemicals, fix broken machinery, patch leaking valves in a tank, etc.

Tanks: The estimated quantities of grapes of each variety reflects the needs of individual tanks volumes, and the total tank volume needed for the coming vintage. During fermentation, white varieties need about 95% of the tank volume, while red fermenting varieties need about 80%–85% of the tank volume to allow some volume for the cap.

After fermentation and skin separation of reds, the must volume reduces to about 60%–70% of the original volume. So, the volume needed for fermentation is greater than storage volume by some 20%–30%. Each fermenting time is short, but the harvest period span is long due to the various varieties maturation dates. So, careful calculation is needed to plan the tanks available and the planed crushing scheme. Also, some of the tanks may still be full with wine from the previous year.

New barrels may be needed and extensive bottling may be required to ensure enough empty tanks for the new coming wine. Each tank should be checked for corrosion, valves defects and any leakage problems.

If some barrel fermentation is planned for white wines, the barrels should be new barrels, checked with water for possible leakage. The number of barrels should be carefully matched to the quantity of wine planned to ferment. Each 225L barrel will contain the must of about 400 Kg of grapes with 5% of overhead space needed during fermentation. A clean, cool space should be reserved for this purpose.

Cooling system: The cooling unit should be in good mechanical condition, including the cooling liquid (usually ethylene glycol), which has to be checked for proper operation in the whole system. All pipelines from the cooling units to the tanks and heat exchangers should be leak-free and thermally isolated. It is highly recommended to test the whole system for its total cooling capability, and every individual tank for its thermostatic control operation. Detailed inspection of the cooling system might save a lot of problems later during the wine processing.

Machines: All processing machines, such as scales, hoppers, conveyers, crushers/destemmers, presses, pumps, lees filters, hot water pressure machines, and forklifts should be in good working condition. The machines should be checked for any potential mechanical or electrical problems and appropriate measures should be taken. The hoses should be checked for leaks and enough hoses of different sizes and lengths should be ready for use.

Chemicals: All chemicals used in the winemaking process should be at the winery, each one according to the amount needed with some to spare. The chemicals to be prepared are:

- Sulfur dioxide (as liquid in steel tanks, or as powdered potassium-meta-bisulfite).
- Tartaric acids.
- Yeast in various strains needed.

- Malo-lactic bacteria (culture or dry).
- Yeast nutrient.
- Di-ammonium-phosphate.
- Fining agents (bentonite and any other needed).
- Filter pads of needed sizes, and diatomaceous-earth (DE).
- Cleaning and sanitary materials.

Laboratory equipment and chemicals: The laboratory should be ready for must and wine analysis. All the equipment for the basic analysis of Brix, total acidity, pH, color intensity, volatile acidity, sulfur dioxide, alcohol, malo-lactic fermentation, and residual sugar should be ready including calibrated solutions. All chemicals needed for the general laboratory work should be on hand.

Sanitation: A clean winery may prevent some microbiological troubles during the wine processing. Any piece of equipment, which will be in contact with the must or wine (machines, tanks, hoses, barrels) should be cleaned with suitable cleansing agents, and rinsed with plenty of water. The whole floor area of the winery should be cleaned with a hot-water pressure machine.

The sewage system of the winery must be checked for possible blockage or leakage. If there is any risk of shortage in water supply during harvest time, it is highly recommended to have a full tank of water on reserve. A winery with a sudden water shortage faces a very serious problem.

CHAPTER II

HARVEST

Chapter ii: Harvest

A. Destemming & Crushing

The purpose of destemming is to separate the stems from the must, as they contain very high levels of tannins, and may contribute a hard "vegetal" or "green" flavor to the wine, if the stems are fermented with the must.

The purpose of crushing is to break the skin of the berries, which releases the juice. These two operations of destemming and crushing are usually done by one machine. In some machines, the destemming is done first, and then the crushing, in some others, vice-versa. The process differs for white and red grapes. We prefer the first method (destemming/crushing), because by doing so less stem material enters the must.

1. White grapes

Fermentation of white must is not carried out with the skins (unlike the red). Therefore, the separation of the stems and the skins from the juice is a necessity. The destemmer/crusher, followed by press, is not the only method to do this. Direct pressing the whole clusters without destemming and crushing can also do it. The idea behind pressing the whole clusters is to minimize skin contact within the must. This technique is most practical in the processing of sparkling wine and in grape varieties that have some bitterness in the skin, such as Muscat-related varieties.

During this mechanical treatment of the must, the absorption of oxygen is very intensive due to the large surface area that the must is exposed to. The oxidation (and hence browning) of white must is very fast, especially if the must temperature is not low enough (above 10°C) and is considered detrimental to the quality of the wine.

The time taken for mechanical processing of the grapes, especially whites, should therefore be minimal. The grapes are crushed and sulfur dioxide is added (see in General Aspects chapter) as soon as possible. Recently, some wineries do not add sulfur dioxide at this stage (in white wine), in order to oxidize and polymerize some of the phenols, which may then precipitate during fermentation and fining. By doing so, the later browning of the white wine may be reduced. This early (intended) oxidation of the white must, in contrast to the attempts to prevent it, is somewhat controversial. After fermentation, during wine processing operations, the exposure of the wine to air by racking, pumping leaks or filtering, causes additional oxidation, which may damage the wine's quality.

In certain cases where skin contact in the must is desired, the crushed must can be transferred through heat exchangers in order to lower the temperature to about 10°C, and then to a drainer or to a storage tank, for some hours, before pressing. If skin contact is not desired (in most cases), the must is transferred through a heat exchanger directly to the press, and then to a settling tank.

2. Red grapes

After destemming/crushing, the must is transferred to the fermentation tank to begin fermentation. Separation of the skins from the must is done later, during or after fermentation, according to the winemaker's decision.

If "blush wine" (blanc de noir) is to be made from red grapes, the must is transferred from the destemmer/crusher through the heat exchanger and to the press. The free-run and the press-run are generally separated and blended later, according to the desired pink color of that wine. From the press the juice is pumped to the settling tank to be treated as white wine.

The red skins of the blush juice, coming out of the press, can be added to the must of some other red wine (in fermentation stage), to enhance color, tannin and body, if desired.

In certain cases the stems are used to enhance the tannin level and to add complexity to the wine. This method is sometimes used with Pinot Noir grapes, which might lack color and tannin in its skin. The must containing the stems is blended later with the rest of the wine. The stemmed portion may vary between 20% and 40%, according to the case.

B. Skin Contact

Skin contact is one of the most important aspects in the winemaking process. It influences the type, character, aging period and general quality of the wine. Some of the varietal flavor and most of the color and tannin compounds in grapes are stored in the skin. The total amounts of phenolic compounds vary in ranges of 2–5 gr/kg in red grapes, and 0.2–0.5 gr/kg in white grapes.

For details on the chemical structure of the phenolic compounds, and more details on their function, see the General Aspects chapter, section D. Roughly, they may be divided into three functional groups, responsible for *color* (pigments), *astringency* (tannins), and *flavor* (bitterness). The pigments contain mostly anthocyanins (cyanidin, peonidin, malvidin, petunidin, and delphinidin), which give the wine its red color. The color is pH dependent and changes from deep red at low pH to green-blue at about pH 5-6, with no sharp color transformation. They are generally bound to sugar molecules as glucosides.

Tannins are the polymeric forms of the phenolic compounds such as catechin, leuco-cyanidin, gallic acid, vanillic acid and others, and also of anthocyanins. Most polymers are composed of two and up to six or seven monomers, containing many OH sites, which are responsible for the "dry" feeling or astringency of red wine.

The third group, the flavor compounds, is mainly derived from cinnamic acid derivatives (caffeic acid, coumaric acid, ferulic acid), and partly from the flavanols group. The above functional grouping is not strictly diverse, and some compounds may be colored and tannic, or tannic and flavored at the same time. The intensity and dispersion of these compounds in the skin depends upon many factors, such as variety, maturity state, climate, soil and cultivars management. This chapter deals with the extraction of these compounds from the skins into the must.

During the vinification process, the extraction rate of the phenolic compounds depends on the variety of the grapes, as well as the temperature, alcohol concentration, sulfur dioxide concentration and time of contact. The extraction is directly related to the last four factors. It is faster at higher temperatures, at higher alcohol and sulfur dioxide concentrations, and its concentration increases with length of contact (maceration). The total amount found in the finished wine is about 1- 2 gr/L in young red wine, and 0.1- 0.4

gr/L in white. In going into detail about skin contact, we will separate the discussion on white and red grapes.

1. White Grapes

In white grapes, the skin contact intensity can be categorized into three groups: no contact, short contact and long contact.

In *no-contact treatment*, the must is pressed immediately after destemming and crushing. The most extreme practice of this minimal skin contact is achieved by pressing the whole clusters without crushing. This technique is used mostly in sparkling wine production, where neutral basic wine with minimal character is desired.

Short skin contact lasts between 1 to 4 hours, where *long skin contact* can last for up to 24 hours. In these cases, the must is left with the skins after crushing, in order to extract more varietal aroma, knowing that there is also a gain of some undesirable bitterness from the flavonoid compounds, which later will be oxidized and will turn the wine color to yellow-brown.

When skin contact is desired, the must should be sulfited right after crushing, mixed well, and be cooled to 10°C–15°C. The major effect of temperature on skin contact is the increasing rate of phenolic compounds extraction with increasing temperature. On the other hand, lowering the temperature at this stage is necessary in order to reduce the oxidation potential of the must.

The length of skin contact depends on the variety, the ripeness of the grapes, and the kind of wine to be made. Some winemakers prefer minimal skin contact to get fresher, non-astringent white wine, while others prefer more varietal character, richness in aroma, with certain tannin levels.

For aging white wine, it is better to have some skin contact, which, through aging, the phenolic compounds will contribute to the bouquet of the wine. When the grapes are fully ripened, the length of skin contact can be shorter. However, if they are not ripe enough and skin contact is longer, care should be taken, as the must and the wine may become "green" or get a "leafy" flavor, which may be undesireable.

After skin contact, followed by pressing, the lees must be separated. This last operation can be done either by gravitational settling for 12 to 24 hours, or by centrifugation. In the gravitational settling, the lees settle down in the tank to form a dense layer whose thickness depends on the pressing technique used. The clear must is then racked off from the lees layer to start the fermentation. The lees left over can be filtered by a lees filter and be added to the rest of the must. Centrifugation, on the other hand, can reduce the solid content of the must to about 1% solids at a very fast rate. The musts flow through the centrifuge is continuous, with interruptions from time to time to discharge the solids. The centrifuge can also be used at any other stage of processing, primarily after fermentation, to clear off the wine from the yeast lees. When centrifuging wine, care should be taken to prevent air from coming into the centrifuge (by using a nitrogen environment), because this process introduces a lot of gas into the liquid.

A somewhat different technique to treat white must is to use a drainer (juice separator). The drainer is a vertical tank that contains an inner screen (central or on its side), which the clear juice can flow through, leaving the skins behind. The drainer is filled with must directly from the crusher or through the heat exchanger, and the must is left in for a certain period, according to the desired level of skin contact. The juice is separated from the skins through the screen, either by gravitational pressure of the must, or by CO_2 pressure (up to 7 p.s.i.) exerted from a CO_2 tank. The principle of the drainer is that the pomace itself functions as a "filtering" media, reducing the solid particle content of the free-run juice. After the free-run juice has run out to another tank, the rest of the must is discharged into a press to extract the remaining juice. The pressed juice is combined with the drained juice in the settling tank, where the total solids content is lower compared with a must that has not been drained. The whole process of draining (loading, waiting, draining the free-run and discharging the must into the press) may take from about one hour, when minimal skin contact is desired, up to many hours as needed in a long skin contact.

Usually two or more drainers can serve one press. The drainers also function as a storage buffer between the destemmer/crusher and the press. The drained juice contains about 0.5–2% solids, compared to 3–8% in the usual press separation, which needs a settling time to reduce this high solids content.

A remark is made here concerning the Muscat varieties (such as Muscat of Alexandria, Muscat Canelli, Sylvaner, Symphony, Emerald Riesling and others), which may develop a typical bitterness in their wines. Minimal skin contact may reduce this problem. Pressing the whole clusters without crushing may be the best method for such cases. On the other hand, in Chardonnay, skin contact is generally practiced, where in other varieties it depends on the case and on the wine's style.

2. Red Grapes

The typical extraction rate of the color and tannin from the skins can be seen in the following figure:

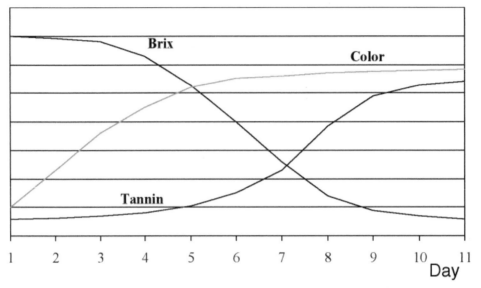

Schematic diagram of the extraction of color and tannin from the skins during and after fermentation.

The color pigments, which are mainly monomeric anthocyanins, are extracted faster than the momomeric and polymeric tannin molecules. At around zero Brix there is no more color extraction, therefore there is no need for additional skin contact. Tannins, on the other hand, continue to be extracted as long as the skins are in contact with the wine (until equilibrium is reached between the skin tissues and the external liquid, the wine). Therefore, separation of the skins from the must can be done at any time along the extraction line, depending on the grapes and the expectation of the wine. If the separation is done shortly after the beginning of fermentation, the result is a good red color with minimal amount of tannins, and the wine may be consumed very young. If separation of the skins is made after a long period of contact (long after the fermentation had been finished), the tannin extraction will be high. The wine likely (depending on the case) will be very tannic, hard, and will require longer aging.

In red grapes, we can point out three kinds of skin contact treatment, which will reflect the type of wine being made:

a. No skin contact: immediate separation of the skin from the juice. This wine, which may be almost white or slightly pink in color, will be called "blanc de noir" or "blush" wine. This wine is considered white wine, and its processing is carried out accordingly.

The blush wine, which became very popular during the "eighties" in the United States, is a light, fresh, fruity wine, usually off-dry, with about 1%–3% residual sugar. The main problem with blush wine is that the bright pink color ages in a short time (in certain cases in months) to become brown-pink, which is not appealing to the customer. The varieties that suffer the least from this problem are Cabernet Sauvignon and Zinfandel. A possible solution to minimize this color change in other varieties is to avoid the use of SO_2 at crushing before fermentation, allowing part of the pigments to oxidize and polymerize, which then settle down later during and after fermentation. At bottling, with enough free SO_2, the color of that wine will be more stable.

b. Short skin contact: before fermentation. The contact may last a couple of hours, up to about 24 hours. This "blush" wine may also be called "Rosé," and the intensity of its pink color depends on the skin-contact length. The longer the contact, the more intense the color. After press, to separate the skins, the must is processed as white wine to become a rosé wine.

c. Long skin contact: during fermentation. This is the classic red wine production, where the skin contact is going on during fermentation and its duration may last from a minimum of 3 to 4 days for light red wine, up to the common practice of 14 to 21 days, and in certain cases even for 4 – 5 weeks in heavy, tannic, long aging wines. The variety, the specific grapes, the style of the wine, and the expected aging time are factors that determine the length of the skin contact during and after fermentation.

Some grape varieties are intense in color and tannins (e.g. Carignane, Cabernet Sauvignon, Zinfandel, Petite Sirah and others), while other varieties (especially Pinot Noir) have light pigments and low tannin concentration. The color intensity also varies from season to season, and the decision when to separate the skins has to be made in accordance with each wine. The tannins are colorless, but they may contribute to the color intensity by co-polimerisation with the red colored anthocyanins to produce co-polymers (anthocyanin-tannin), which are more intensely red colored then the anthocyanin itself. So, the desired tannin level in wine (responsible for the wine astringency and bitterness) should be considered not only in prospects for aging, but also in its color intensity effect. For example, in Pinot Noir wines, which have low tannin levels, some wineries ferment certain portions of the must with the stems (which contain a high tannin level), and blend this portion with the rest. The blend is higher in tannins and somewhat more complex (and 'green') but has a better chance for higher color intensity.

One of the methods to enhance low color intensity is *thermovinification*, namely pre-heating the grapes or must, prior to fermentation. The heat destroys the skins cells membranes, releasing the pigments, tannins and other substances into the must. Therefore, the yield of extraction is temperature dependent, within the practical range of 40°C–80°C. Thermovinification can be done either by heating the must after crushing, or by heating the grapes (whole clusters) before crushing (by hot air or steam). The second method has the advantage of heating mainly the skins with minimal heating of the pulp inside the berries. For wines that are to be consumed young, and when the color intensity is a problem, thermovinification may help to enhance the color. For long-aging wines, the extra color extracted by heating will polymerize quite fast, to high polymers, and precipitate after a relatively short time, yielding no color advantage after several years of bottle aging.

A better method to enhance the color intensity in wine may be to blend it with a small portion of high color intensity varieties, such as Zinfandel, Alicante, Petite Sirah, and others.

On the opposite side of heavy high tannin wines are the light, fresh and young consumed wines. These wines are produced from the appropriate grape varieties, and with a short skin contact treatment, to minimize tannin content and still having a good red color. Most light red wine production is done by a different technique called 'carbonic maceration'. This is a special fermentation technique, which utilizes the ability of enzymes naturally present in grapes to transform some small amount of sugar into ethanol. The process is eventually stopped by the accumulating alcohol inside the berries, which poisons the enzymes in the cells at about 2% ethanol. CO_2 is also formed in this intracellular 'auto fermentation' (without any microbial activity). The enzymatic activity may be enhanced at high levels of CO_2 (over 50%), or at low oxygen levels (below 5%). This technique was developed and is practiced mainly in France, in the production of light red wines with low tannin, primarily fresh-style wines intended to be consumed young (*Beaujolais Nouveau* wine). In this technique the grapes are uncrushed, and the whole berries (with stems) are placed in a tank filled with CO_2 gas to decrease oxygen contact during the process. This is the origin of the name 'Carbonic maceration'. The berries are partially crushed by the clusters weight and juice flows to the bottom of the tank, where natural yeast fermentation takes place. After several days (up to a week or two) the grapes are pressed and left to finish theirs yeast fermentation to dryness. By such vinification one can get good colored, softer red wine, with characteristic fresh aroma, which is caused by a series of volatile aldehyde compounds formed during this technique.

To summarize this subject, the very general principles should be remembered: the pigments are extracted more quickly than the tannins, the must should be checked during fermentation and differentiation should be made between the color intensity and the tannin level in accordance with the kind of wine desired.

C. Free-run & Press-run

1. General View

Pressing of grapes is done on the unfermented must for white wine and "blush wine" production, and during (or after) fermentation of red wines. The juice (or wine), which comes out of the press by using no pressure or just minimal pressure, is called *free-run*, whereas the pressed juice (or wine) is called *press-run*. The unfermented free-run must is about 80%–90% of the total *extractable juice*, where the fermented free-run wine is somewhat more, about 90%–95% of the total wine volume. This is because during fermentation the pulp cells die, and the cells' membranes are not active anymore, making for an easier flow of the cells content.

The constituent content of the free and pressed run is different. In the press-run (of must or wine), there are more phenolic and polyphenolic compounds (pigments and tannins), less total acidity, higher potassium concentration and therefore higher pH. In pressed wine, there is also higher volatile acidity and higher "vegetal" or "green" aroma. All these parameters are related to the pressure level exerted during pressing and the type of press used. Based on all parameters, better quality wine is made from free-run, where the use of the press-run has to be considered according to its particular quality. It is advisable to separate the press-run from the rest, and process it in a different tank. If it has good quality, it can be blended later with the free-run wine in such a proportion as to get the highest quality possible. The press-run also allows the winemaker one more possible dimension to enrich his wine in color (in red wine), tannins, and complexity.

Grape juice (as many other fruit juices) contains *pectin*. These polymeric compounds tend to create colloid coagulation in the wine, which may contribute later on as an instability factor. Pressed juice contains more pectin that does free run juice. Hydrolyzation of the pectin during juice processing will prevent this potential problem. The hydrolyzation can be done by pectolytic enzymes, which are commercially available. Wine or grape juice treated by *pectolytic enzymes* will clarify faster and will be easier fined. Its use is recommended for either white or red wines. The enzymes are tolerable to must conditions. Its pH optimum activity is between 3.5 and 5.0. Its optimum temperature activity is at 50°C, but it is active as well between 5°C–60°C. Sulfur dioxide is an inhibitor to its activity, but it can tolerate SO_2 concentration up to 500 ppm. The enzyme is also tolerable to all wines' alcohol levels, so it can also be used for finished wine. The practical range of enzyme addition is 2.5–5.0 gr/HL of must.

2. Presses

Presses can be divided into two major categories: batch presses and continuous presses. In the first category, the oldest and simplest press is the vertical basket press, usually made of wood. The pressure (up to 5 kg/cm^2) is exerted either manually by a screw jack or mechanically (hydraulic, pneumatic, or electrically), compressing the pomace downwards.

By pressing the pomace, the juice runs out through the basket rim and is collected from its base. The basic problem with this kind of press is that the pressure is exerted on the pomace block, leaving some of the liquid trapped inside with no way out. The pressed "cake" is always wet after pressing. In other word, the pressing yield is low. Also, unloading the pomace after pressing, namely, breaking the compressed "cake," is difficult and requires hard work.

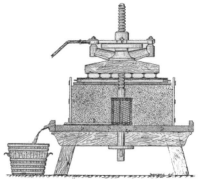

Old vertical basket press

A modification of this kind of press is a horizontal press, made of stainless steel. The pressure is exerted by two plates moving from the edges of the cylinder, pushing the pomace toward the center and compressing it. The pressure is released from time to time, while the cylinder rotates slowly to break down the pomace "cake," enabling more homogeneous pressure, and allowing easier releasing of the trapped liquid.

A much more sophisticated and modern version of this press is based on inflating an air sac (bladder) mounted to the side of the cylinder. In the older type of inflating press, the air sac is centered in the drum, pushing the pomace towards the walls. In modern presses of this type, the bladder is mounted to the sidewall of the cylinder. The pressure in the sac is built up gradually in cycles, where each cycle is composed of the following: pressing the must gradually while rotating the drum, holding the pressure at its maximum value and releasing the pressure while reversing the direction of rotation (to break the pomace "cake"), then pressing gradually again while changing back the direction. The pressing time and the reverse rotation time in the cycle are set prior to operation. Normal cycling time is 2–5 minutes. The press operates in three cycling blocks according to the maximum air pressure in the bladder. The number of cycles in each block section is determined by the press operator and is automatically controlled and programmed. In block I, at low pressure (about 0.2 bar), the number of cycles is about 10 to 15. In block II, the pressure is incrementally increased at each cycle by 0.2 bar, up to the maximum pressure, which is pre-set by the operator (about 2 bars). In block III, the pressure in each cycle is set to the maximum, about 5 to 10 cycles.

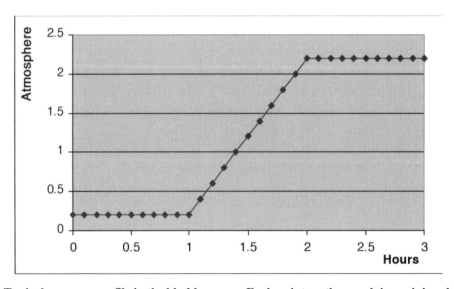

Typical pressure profile in the bladder press. Each point on the graph is a mini cycle.

The total pressing time (including loading, pressing, and discharging) takes about 2–4 hours. Because of the relatively low pressure exerted on the pomace and the gentle mechanical treatment of the skins, the solid content in the juice is relatively low (about 2%–4%).

This is most important in white wines, because it allows one to start the fermentation with relatively clear juice, which is crucial in producing quality white wines. The distinction between the free-run and press juice is determined at a certain pressure level during the pressure slope. Free run juice is determined at a certain low pressure (say 0.4 atm.), and from that point the coming must from the press is considered as press run. Another option as to when to separate the free run from the press juice is by tasting the juice coming out of the press. The parameter for judging is the change in acidity taste. It gradually gets lower (with flat taste).

In a case where the skins are thin and slippery, the screens inside the drum are usually plugged, causing difficulties in the press operation (this phenomena is typical, for example, in Semillon or Sauvignon Blanc). A good practice in such a case is to add a portion of the stems (from the destemmer/crusher machine) to the must when loading the press (layer of must, layer of stems and so on). The existence of the stems prevents plugging of the screens.

All types of presses described above are batch types. They must be loaded and discharged at each batch. The batch quantity may run from about 2 HL (small basket press) up to 7–8 tons in the bigger horizontal air sac presses.

The second category of presses is the continuous press, which is based on an "infinite Archimedes screw" that presses the pomace slowly to an outlet whose size is controlled. This enables the operator to set the exerted pressure on the pomace as desired. The advantage of this type of press is that it is continuous, saving the loading and discharging operation. The main (and serious) disadvantage of the screw-type press is that it causes tearing and breaking of the skins tissue, leading to more solids in the juice. A continuous press is not recommended for pressing white must, if recommended at all.

3. Lees

The lees are the solid particles suspended in the must after crushing and/or pressing. This term is also used to describe the sediment that settles before the first racking or after fining, which contains mainly yeast cells or fining particles. When leaving the white must to settle for 12 to 24 hours before fermentation (depending on the particles size and quantity), the settling lees create a dense layer (sludge) of variable thickness, from 5%–10% of the liquid volume.

The juice above the lees layer can be racked off through the racking valve in the tank, but the only way to save the juice trapped in the lees layer is to extract it by mechanical force. This can be done either by lees filter, or by centrifuge.

A lees filter can be very effective in clearing the juice from its solids. The filter contains 20–40 plates, which support thick cloth pads. The plates with the pads are packed together tightly in a row, at very high mechanical pressure (hydraulic pressure of about 300–500 bars on a 40x40 cm plates). At the entrance of the filter there is a strong pump, which can exert high-pressure forces on the lees to flow through the pads.

To facilitate the filtration of the dense sludge, it is mixed well with diatomaceus-earth (DE) before entering the filter. The DE is a finely ground mineral powder which increases the surface area of the filtering pads. This enables the lees to combine during filtering in a multi-layer texture with the DE, rather than on one layer of the pad surface without the DE. The mixing ratio of DE with juice is 2–4 Kg/HL, depending on the density of the lees. More DE is needed when the lees density is higher. The first batch of mixing can be 3–4 Kg/HL. When this batch is filtered, and a new batch of juice is mixed with DE, the DE/juice ratio can be cut to half of the first one. If a third batch is prepared, the ratio is a quarter of the first addition.

At the beginning of filtering, the flow is easy and no pressure is developed. The first batch of the mixed juice and DE is used to build the DE "cake" between the pads, protecting them from being plugged by the fine lees particles. As the volume of liquid flowing through increases, more DE/lees particles are deposited, increasing the flow resistance. Pressure gradually builds up by the filter-increased resistance. The maximum workable pressure in most filters is about 10 bars. The pump is pressure controlled, and stops when reaching the maximum pressure for a couple of seconds, until the pressure drops and it starts again. When the pumping intervals exceed 10-15 seconds, the filter is probably plugged and filtration should be stopped. The filtering capacity of a lees filter with 40x40 cm plates is about 30–50 liters/plate, so a typical lees filter with 20 plates can filter 600–1,000 liters of must before it will be plugged.

Some useful hints for the lees filtering operation:

- At breakdown of the filter plates, the shape of the "cake" reflects the quality of the filtering. A "good cake" (dry, dense and easy to remove) shows the proper ratio between DE/lees and good pad condition. When the cake is soft (not homogeneous and sticky), it means that the filtering was not efficiently done, the DE/lees ratio was too low, or the cloth pads were almost plugged with lees particles by early use of the filter.
- Good washing of the pads with water after each use is necessary to prevent microorganism growth. Also, after a couple of filterings, the pads get partially clotted, and it is necessary to take them off the plates to machine-wash them.
- Before closing and compressing the plates for new filtering, wet the cloth pads with water.
- Check the edges of the pads before closing the plates, to assure that no bent pad is caught between the plates. If this happens, the filter will leak when the filtering pressure rises.
- The first 20–40 liters of filtered juice has an earthy taste (from the DE). Consider discarding it.
- From time to time, open the high-pressure pump valve to release the DE that accumulates.

And finally some remarks:

In general, the quality of the lees filtered juice or wine is poor. When dealing with the lees left over in wine after racking (yeast lees, fining lees), generally the volume of the lees layer is small, and it is sometimes simply better to drain it off.

Also, the DE dust is not healthy to breathe and it is highly recommended that the operator wear a dust mask while mixing the DE with the must.

D. Must correction

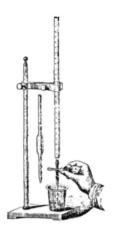

1. Acidity

When we previously discussed the acidity of must and wine, it was noted that the two parameters involved, namely total acidity and pH, are not simply related.

Differentiation should also be made between total acidity and titratable acidity. The *total acidity* is the sum of all organic acids in the wine (or must), including their salts. The *titratable acidity* is the available acidity as free H^+ ions. The total acidity is always higher than the titratable acidity, because part of the organic acids is neutralized by the various cations, mainly potassium. It is clear why the enzymatic measurement of the total acidity will show higher results than the titratable ones. In this book (and in other texts too), the titratable acidity is traditionally called total acidity (TA).

Some words concerning the buffer capacity of must and wines: the buffer capacity, which is the response of the pH changes to addition of strong base (or acid), depends on the total titratable acidity (of the weak organic acids in the case of wine, *and* on the main cation concentration in the media (potassium in the case of wine). The higher the total acidity and potassium concentration, the higher the buffer capacity (with no simple linear relation). A titration curve of a typical must, and of a strong acid for comparison, is shown in the following figure:

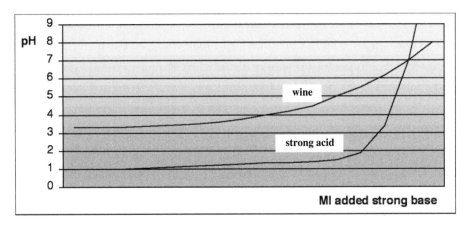

Titration curve of a typical wine (or must) and a strong acid by a strong base.

Note the sharp change in the strong acid titration curve at the neutralization zone, compared with the wine curve, which is quite smooth and continues. The titration end point is of course at pH = 7. When an indicator is used for determining the end point (instead of a pH-meter), it should be the pK_a point of the indicator, which is not necessary at pH = 7. The most used indicator in white wine titration is phenolphthalein, with pK_a value of 8.2. For convenience and common practice, this pH was chosen as the end point of TA measurements in the wine industry. When measuring the TA of red wine, it is impossible to use an indicator because the wine color masks any other color. The end point is therefore determined by a pH-meter, at pH = 8.2.

The buffer capacity of must is in the range of $(4-6) \times 10^{-2}$ equivalent/L per one pH unit, meaning that in order to change the pH of 1L of must by one pH unit, it is necessary to add $(4-6) \times 10^{-2}$ equivalents of strong base or acid (in pH range up to about pH = 5).

Adjusting the pH and the total acidity is recommended because of two reasons:

First: to adjust the sour taste of the wine- it may be either too sour (high acidity) or flat (low acidity). The adjustment is made according to the must at hand and the desired type of wine.

Second: adjusting the wine pH (and acidity) affects its stability to chemical and microbial changes. This case is mostly relevant in wines with too high pH.

The *sour taste* is more dependent on the concentration of acids (*total acidity*) than on the proton ion concentration (expressed by its pH). As a very general guide, the preferred values of the must would lie in the range of 6 – 9 gr/L total acidity, pH = 3.1-3.4 for white wines; and 5–7 gr/L total acidity, pH = 3.3–3.8 for red wines. In sweet or semi-sweet wines, the acidity should be in the higher range, and the pH at the lower one, in order to balance the sweet taste. It should also be kept in mind that the total acidity (and the pH) may change during the wine processing. First, it becomes higher after fermentation by the addition of new acids produced as by-products of the fermentation (e.g. succinic acid, acetic acid, malic acid, and others). Second, the acidity can also be reduced by malo-lactic fermentation and by cold stabilization (see later).

The acidity issue can be divided therefore into two categories: *deficiency* and *surplus* of acid, or in other terms, too *high* pH and too *low* pH.

Too high pH

The more severe and common case in white must is the lack of acid rather than the surplus of it, especially in warm regions. The advantages of correcting the pH and increasing the acidity, when it is in deficiency, are: more controlled and steady fermentation, a reduction of the yeast autolysis at the end of fermentation, and inhibition of bacterial spoilage. It also preserves the flavor and the aroma of white wine, and is necessary for the balance and general quality of white wine. Too high pH is also not desired in red wines, mainly because of microbial stability problems, and because of reduction in color intensity at too high pH. Preferably, the acid correction should be done before starting the fermentation, because of the above reasons regarding fermentation, and because any major correction later on will greatly affect the wines stability.

The techniques used for acid correction when it is deficient are: blending with high acid must, addition of acids, and ion exchanging.

(1) Blending with other must (high acidity and low pH) is the best natural and elegant method, whenever possible. However, in many cases, it is not practical. It may be quite difficult to find the right grapes with the desired pH and total acidity, at exactly the time of crushing and starting fermentation.

There are also other factors that should be considered in blending, which are not necessarily dependent only on the pH and total acidity, especially when blending with another variety. Such blending may also change the quality and character of the wine.

(2) The most practical and frequently used method is acid adjustment. In this case the only change is solely the wine acidity and pH. The options are to use tartaric, malic, citric and fumaric acid; each has its advantages and disadvantages.

    a. Tartaric acid is the strongest of the three (lowest pK_a), and would need less than the others to get the same result. On the other hand, a large part of the acid added will precipitate later during cold stabilization. However, by cold precipitation of the potassium-bitartarate salt, the potassium concentration is reduced, which leads to a lower pH value. If high pH is the major problem of the must, the addition of tartaric acid and precipitating its surplus potassium by cold stabilization will improve the pH value of the wine considerably (for more details see the Cellar Operation chapter, section B).

    b. Malic acid is the principal reason for acid deficiency in must during ripening period by its respiration. The addition of malic acid into the must can be looked at as a way of restoring the original balance between tartaric and malic acids in the grapes. Malic acid, which is less acidic than tartaric acid, will not precipitate in cold temperature, but may be transformed into lactic acid by malo-lactic fermentation. When the commercial DL-malic acid is used to acidify the must, only the L-malic acid isomer is consumed during the malo-lactic fermentation, leaving the D-malic isomer (50% of the added DL acid) untouched. So, even though malo-lactic fermentation takes place, the addition of malic acid will partially contribute to the total acidity.

    c. Citric acid can be added to expand the acid taste and to prevent possible iron haze (by complexing the iron). However, during ML-fermentation, it may be partially converted into acetic acid. So, addition of citric acid can be done safely only when ML-fermentation is inhibited (because it isn't desired in that particular wine) and measures have been taken to prevent it. Otherwise, it can be added after the ML-fermentation is completed. There is legal limit in the U.S. of its content in wine of 0.7gr/L, which limits the use of this acid to small quantities.

d. Fumaric acid is another option that can be used for acidification. Its pK_a is practically the same as tartaric acid. This acid is also very efficient in inhibition of ML-fermentation at concentrations above 300-500 mg/L. The main problem with fumaric acid is its low solubility in water or water/alcohol solutions. The solubility in water is 5 gr/L, 23 gr/L and 98 gr/L in 20°C, 60°C and 100°C respectively. The best way to use this acid is to prepare a hot solution of about 50–80 gr/L and add it hot into the must or wine. The exact volume of the solution must be calculated according to its concentration, the must volume, and the desired concentration of the acid in the wine. The legal limit of fumaric acid in the US is 2.4 gr/L. Fumaric acid has some harsh taste and its addition should be done with care.

The amount of acid needed to correct the acidity deficiency depends on the total acidity, the pH and the buffer capacity of the must. In most cases, when using tartaric acid at a pH range of 3.6 ± 0.3, every 1 gram/liter added (0.1% TA), will lower the pH by about 0.1 ± 0.03 pH units. The (-) sign (less then 0.1pH change), is at about pH < 3.5 and the (+) sign (more then 0.1pH change), at pH > 3.5. For malic or citric acid, each 1 gram/liter added will lower the pH by about 0.08 ± 0.02 pH units, at the same pH range as above. This is a general guideline. Before adding the acid to the must (or wine), one should try to determine the right quantity per liter in the laboratory, in order to get the optimal TA and pH according to the case.

In cases where the pH is too high, and the TA is also high, (caused mainly by high potassium content), addition of more acid in order to lower the pH will result in a very harsh taste. This case is difficult to correct. One may consider using phosphoric acid in order to lower the pH without adding too much to the total acidity. The use of phosphoric acid is a common practice in the food industry, although it is not used in the wine industry. It gives the must and the wine a "watery" or low body feeling. Its use is illegal in the US as a wine additive.

(3) The ion exchange method exchanges the potassium cation with hydrogen, and is a good technical solution to the above case of high pH and high TA. The ensuing exchange is very efficient, and almost all the potassium is replaced by hydrogen ions, lowering the pH substantially. So, only a portion of the wine is transferred through the ion exchanger column and blended with the rest of the wine. The result of such blending will lower the pH of the total wine batch.

Too low pH

Deacidification of the must is the other side of the problem, where the must acidity is too high (above 9 gr/L) and the pH is too low (below $3.0 - 3.1$). This problem is usually the result of unripe grapes that had to be harvested because of some viticulture difficulties, such as weather conditions, vine diseases or bunch rot. The problem is quite common in Germany, parts of France, Switzerland and other cool regions. The taste of the wine made from such must is probably sour, tart and unbalanced. The low pH will also inhibit ML-fermentation when it is most desired to reduce the acidity. In order to correct the excess acidity and to bring the pH higher, some measures can be taken. Namely, blending with low acidity must, chemical treatment, malolactic fermentation and cold stabilization.

(1) The best method is to blend with low acidity and high pH must. Again, the same consideration mentioned earlier in this chapter on low acidity correction by blending is applicable here as well.

(2) The other way to reduce the acidity of the must (or wine) is by using chemicals. In principle, the neutralization of the excess acid by base is not good, because the salts formed will give the wine a salty taste and sometimes an off-flavor. It is preferable to remove the excess acid by precipitation with calcium carbonate, potassium carbonate, or potassium tartarate. All will react with the tartaric acid to produce hard soluble tartarate salts, which will precipitate, and lower the TA:

$$CaCO_3 + 2H^+ + 2HT^- ===> Ca(HT)_2 + CO_2 + H_2O$$
$$K_2CO_3 + 2H^+ + 2HT^- ===> 2KHT + CO_2 + H_2O$$
$$K_2T + H^+ + HT^- \qquad ===> 2KHT$$

The reaction can also include the second dissociated anion of tartaric acid (T) to give CaT.

Also, insoluble double salts of tartaric and malic acids can be formed.

The quantities of carbonate or tartarate, which have to be added, must be determined in the laboratory before addition. As a general guidance, experience shows that the following quantities of deacidification agents, which upon addition to the must will increase the pH value by 0.1 pH unit, are:

$$CaCO_3 \qquad : 0.3–0.4 \text{ gr/L}$$
$$K_2CO_3 \qquad : 0.4–0.5 \text{ gr/L}$$
$$K_2T \qquad : 1.5 \text{ gr/L}$$

It is important to remember that although the salts formed are insoluble and precipitate, the Ca^{++} or K^+ ions concentration in the deacidified must or wine is increased, sometimes to the point where instability problems may arise.

(3) ML-fermentation is a good and "natural" treatment for reducing acidity in wine. In this fermentation, the malic di-acid is transformed into the lactic mono-acid, consequently reducing the concentration of wine acidity:

$$HOOC - CH_2 – CHOH – COOH =====> CH_3 –CHOH – COOH + CO_2$$

Also the pK_a of lactic acid is higher than the pK_a of malic acid, which mean that lactic acid is weaker then malic (besides reducing the malic acidity by half).

A major problem in using this method is that when the pH is below 3.1–3.2, it is difficult to carry on the ML-fermentation. In such a case where the pH is lower than this value, it can be brought up to above 3.2 – 3.3 by other methods (such as carbonate addition, or blending), and inoculated with ML culture.

(4) Cold stabilization is an important operation during the process of winemaking. In this operation, the excess tartaric acid is precipitated as potassium-bitartarate salt, lowering the pH (because of the potassium precipitation). So cold stabilization reduces the total acidity on one hand, and reduces the pH on the other. This controversial aspect of cold stabilization on deacidification has to be considered carefully in each actual case. It may, in fact, be better to cold stabilize the wine first, before other methods of reducing acidity are used.

2. Sugar

There are several reasons for the deficiency of sugar concentration in the grapes at harvest time, and unripe grapes cause all of them. When this happens, leaving the grapes on the vine for further maturation and delaying the harvest may not improve the sugar accumulation. In certain cases it may actually cause damage to the grapes. With unripe grapes, there is not enough sugar to ferment into the normal alcohol concentration range in wine. Usually such grapes also contain excess acidity. There is nothing wrong from the quality aspect, to make sugar or acid corrections. However, because these grapes are unripe, the quality may suffer from some more severe problem, which cannot be corrected. Specifically, not enough varietal flavor and aroma. In cold regions (Germany, North of France, Switzerland), the addition of sugar is necessary in certain seasons, and is legal to do. The regulations usually require a certain minimum of natural sugar concentration (around $15 B°$-$16 B°$), notification to the authorities about the sugaring (chaptalization), and notification of the amount of sugaring. In other regions (e.g. in California) it is absolutely forbidden.

The sugar addition can be done with cane sugar (sucrose) or with grape concentrate (about 70% grape sugar). When sucrose is added, it is naturally hydrolyzed to mono-sugars before fermentation by inversion enzymes contained in the must.

The quantity to be added depends on the sugar level in the must and on the desired alcohol concentration. The following formulae are useful in calculating the amount of sugar needed in each unit of sugar terms:

0.92 Kg sugar/HL ====> 1 Brix

1.65 Kg sugar/HL ====> 1 Baumé

2.10 Kg sugar/HL ====> 10 Oechsle

For example, 0.92 Kg of sugar added to 1 HL (100 L) of must will increase the Brix by 1 unit.

If cane sugar is used, it is best to dissolve it in boiling water to make a very highly concentrated syrup before adding to the must. The volume needed can be calculated from the syrup concentration.

To prepare concentrate syrup, boil 25 liters of water, and add 50 Kg of sugar while mixing. After a few minutes the syrup will clear up. The syrup volume will be about 55 liters at roughly 66 Brix.

To measure its exact concentration, dilute a sample by 5 (1 volume of syrup and 4 volume of water), and measure its Brix with a 16B°- 24B° range hydrometer. The syrup concentration will be the Brix reading multiplied by the dilution factor (x5).

Knowing the Brix unit is grams of solid dissolved in 100 grams of solution, the volume that has to be added can be calculated by the following example. Assume that the syrup is 66B° and you want to add 3 more Brix to the must. Then add: 3x92/66 = 4.2 liter of syrup/HL of must.

It is recommended to add the syrup while it is still warm (40°C–50°C), before the sugar recrystallizes. It should also be noted that some dilution of the must with water is caused by this operation. If grapes concentrate is used, the producer of the concentrate indicates the sugar concentration, and the same calculation can be made.

The addition of sugar has to be done at the beginning of fermentation, when the yeast is vital and highly active. If added at the end of fermentation, it may sometimes get stuck and will be difficult to achieve dryness.

E. Cooling and Temperature Control

The temperature is one of the external circumstances which has the greatest share in influencing the act of fermentation. It has been considered, that a heat of about the 54° of Fahrenheit's scale is that which is most favorable to this process. There is nevertheless some latitude to be allowed; but in a temperature either very cold or very hot it does not take place at all.
—From Remarks on *The Art of Making Wine*, J. Macculloch, 1829

1. Fermentation temperature

Fermentation is a heat source process. Yeast ferment sugar in order to get their energy out of it. This aerobic process has a poor yield of energy, and the yeast uses about 57% of the available energy in the transformation of sugar into ethanol. The rest is released mainly as heat (and some as mechanical work of the CO_2 bubbles). Theoretical calculations show that each 1B° releases, by fermentation, about 1.15Kcal as heat. For example, 1 liter of 23B° must will release 26.5 Kcal. In other words, if the fermenting tank is thermally isolated without releasing its heat, the temperature of the fermenting wine could rise (theoretically) by 26.5°C from its starting temperature. Practically, this is not the case. First, if the temperature rises to above 36°C, the fermentation will stop. Second, natural heat losses through the container's walls reduce the expected theoretical temperature. This reduction depends on the total volume of the must (or more accurate, on the *walls surface/volume ratio*), on the container material (cement, wood or stainless steel), and on the environmental temperature (temperature difference between the fermenting must and the surroundings). The natural heat loss is not enough in most conditions, and cooling is necessary. Therefore, controlling the temperature during fermentation is one of the most important factors in good winemaking. The range of temperatures in which yeast is well active and ferments is between 10°C–35°C (50°F–95°F). At the higher temperature range, the fermentation starts faster, but as the alcohol concentration increases, it slows down. At around 35°C it may even stop, leaving some residual sugar unfermented. At low to moderate temperatures, the fermentation starts slowly, proceeds more moderately, and generally will go to dryness. The time lapse between inoculation and the first signs of fermentation (lag period) can take from a few hours to a few days, depending on the temperature and other factors (which will be discussed later). The alcohol formed during fermentation is an inhibitor for yeast growth, and its inhibition effect is greater at high temperatures. It also evaporates more through the CO_2 bubbles at higher temperatures, which makes the alcohol yield higher at low temperature fermentation. The variation in alcohol content can be as much as 1% absolute alcohol difference if the fermentation is carried on at temperatures of 20°C and 10°C. Also, at lower temperatures, the fruitiness of the grapes is better preserved, by reducing evaporation of volatile aroma components from the must. Also, the volatile acidity level at low temperature fermentation has been found to be lower than at high temperatures.

For all the above reasons, and for practical considerations, it has been accepted that the preferred fermentation temperature for white wines is between 8°C–14°C (46°F–57°F). This is also true for rosé wines, and white wines made from red grapes ("blush wine"), as they are considered white wines. For red wines, on the other hand, the fermentation temperature should be higher, between 22°C–30°C (72°F–86°F), because of two reasons: better color and tannin extraction, and because less fruitiness is desired in red wines.

2. Temperature Control

In order to carry on the fermentation at the desired temperature, one should control the temperature by means of a mechanical cooling system. This modern technique uses cooling liquid (usually ethylene glycol) flowing in a cooling jacket around the tank. The liquid is cooled by a cooling system, which must have the cooling capacity needed for all winery functions. The tank's jacket is generally built as one or two strips at 2/3 and 1/3 of the stainless steel tank's height. The fermentation heat flows from the tank's walls to the cooling liquid, and then away to the cooling machine. The temperature can be controlled individually at each stainless steel tank by a temperature sensor placed in the tank. The sensor should be long enough so it will read the liquid temperature far away from the walls, where the temperature is the coldest.

Because there are two temperature gradients in the tank, one radial, from the center toward the walls, and the second from top to the bottom of the tank, the temperature control does not necessarily control the whole volume at the same temperature. During fermentation these gradients are reduced by the fermentation turbulence, and by the pumping-over of the must during red wine fermentation. In old European wineries, the fermentation cooling is done simply by dripping cold water on the outside of the tank's walls, with the flow rate of the water serving as the temperature control. Another method is to use an air-conditioned room, which is much less efficient, because of the low heat capacity of air. In barrel fermentation (of white wine), this is the only method to control the fermentation temperature.

In red wines, the fermentation may take between one and two weeks, whereas in whites, it may take three to six weeks to finish. In white wine, if the fermentation seems to be slowing down or even stuck towards the end of the process, it is sometimes advisable to raise the temperature to $15°C–18°C$ in order to complete the fermentation.

After fermentation is over, and during the rest of the winemaking process (racking, fining, barrel, etc.) up to bottling, the temperature should be controlled at $20°C \pm 5°C$ for reds (preferably at the lower range), and $12°C$ $–15°C$ for whites.

When quick cooling is needed (in white must), such as when the must has to stay on the skins for several hours at a cool temperature, there are cooling machines that allow one to reduce the must temperature by $5°C–15°C$ at a very high flow rate through a heat exchanger. In any fast cooling with a heat exchanger, it is highly recommended to pass the must through the cooling system to another tank, rather than to pump it back to the same tank. By returning the cooled liquid back into the same tank, the temperature gradient between the cooling system and the liquid gradually gets smaller and smaller, with lower and lower cooling yield. When the liquid is transferred to another tank, the gradient remains constant, making the cooling much quicker and more efficient.

Fermentation

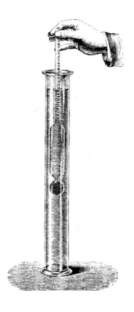

Chapter iii: Fermentation

A. General view

Fermentation is the heart of winemaking. The goal in this process is to convert sugar into ethanol in a way that produces a minimum of undesired byproducts, while preserving the maximum natural aroma and flavor of the fruit. Ideally, fermentation produces ethanol in the desired range, while enhancing the unique character of the alcoholic beverage.

The basic expression of the fermentation process is given by the overall formulation:

$$C_6H_{12}O_6 \longrightarrow 2CH_3-CH_2-OH + 2CO_2$$
$$MW = 180 \qquad\qquad MW = 46$$

Theoretically 180 grams of sugar will produce 92 grams of ethanol, which is 51% of the sugar weight. Actually about 5% of the sugar is consumed to produce byproducts such as glycerol, succininc acid, lactic acid, 2,3-butanediol, acetic acid and other products. Also, about 2.5% of the sugar is consumed by the yeast as a carbon source, and about 0.5% is left over as unfermented residual sugars. Consequently, a total of about 8% is not converted into ethanol. This makes the weight yield: 51 x 0.92 = 47%, which is 180 x 0.47 = 84.6 grams of ethanol. The volume of that amount of ethanol, produced by a mole of sugar (ethanol density at 20°C is 0.789 gr/ml) is
84.6/0.789 = 107.2 ml. When this volume of pure ethanol is mixed with water to make 1L of solution, the total volume contracts by 0.7% (at around 10% alcohol). So, the alcohol concentration of wine made from 180 gr/L sugar, would contain 107.2 x 1.007 =108 ml/L of alcohol, which is 10.8% (v/v). In Brix units, (where about 95% of the total solids content of must is sugar), in order to produce 1% (v/v) of alcohol, the following brix will be needed: (180 /0.95 /108 = 1.75), namely, each

1.75 B° ———> 1% (v/v) alcohol.

This formula which is very useful in predicting the alcohol content in wine, is also expressed as the ratio of % ethanol (v/v) to must's Brix, namely, 1/1.75 = 0.57. The meaning of this number is that the potential alcohol content in a given must is estimated by the quite good approximation:

> **Potential alcohol % (v/v) = 0.57xBrix**

The experimental value of this number was measured in various studies and was found to be in a very good agreement (varying in the range of 0.55–0.60). There may be some limitations to that formulation: (1) The non-sugar solids, which are part of the must's Brix, depend on the grape variety, growing region and state of maturity. Less ripe grapes have higher non-sugar solids, hence a lower alcohol/Brix ratio. (2) The ethanol yield depends on fermentation temperature: the higher the temperature, the lower the yield (partly because of other by-product such as glycerol, and partly by ethanol evaporation).

B. Wine Yeast

People have been making wine for thousands of years without any idea of why or how the grape's sweetness was transformed into alcohol. It wasn't until Pasteur's time, in the mid-nineteenth century, that it was found to be a microbiological process, conducted by yeast. If the must is left to itself, the 'natural' yeast found on the grapes and in the winery will start fermentation. Over many years, through research, trial and error, the more favorable yeasts for wine production were selected. Currently, about 150 yeast species are known to relate to winemaking, categorized into about twenty different *genuses*. The most important yeasts for wine production belong to the *Saccharomyces-cerevisiae* and *Saccharomyces-bayanus* species. These two species (the first name designates the genus and the second the species) are sub classified into different *strains* of yeast, based on their special characteristics for the winemaking process, like:

a. **Tolerance** to different medium conditions, such as temperature, alcohol and sulfur dioxide concentrations. Certain yeast species are sensitive to cold, becoming inactive at low temperatures. This characteristic can be very useful, easily stopping the fermentation process to leave some residual sugar, simply by reducing the temperature. Other species of yeast are tolerant to high alcohol content, and therefore can be used to ferment high Brix must, or to restart fermentation when it gets stuck toward the end. In general, yeast is not very sensitive to SO_2

at the concentrations used in winemaking during fermentation. However, there are some species that are more sensitive to it.

b. Different by-products during fermentation. Some by-products are undesired, such as hydrogen sulfide or volatile-acids. Sulfur dioxide is another by-product in this category.

c. Ability to ferment to the end, leaving low residual sugar.

d. Flocculation capability (colloidal or granular), which affects the yeast sedimentation after fermentation is over, and hence the wine's clarification. This characteristic is important in sparkling wine production (besides other characteristics, dealing with the unique demands of second fermentation).

e. Kinetic of fermentation is an important factor in winemaking. This characteristic is different in various species.

The commercial yeast strains of *Saccharomyces-cerevisiae* (also called *Saccharomyces- ellipsideus*) and *Saccharomyces-bayanus* are sold in dry vacuum packages or as liquid cultures. The loss of activity during storage of the dry yeast is temperature dependent. The activity loss after one year of storage at 20°C and 4°C is about 20% and 5% respectively. The viability of fresh batch is about 20×10^9 cells /gram. The recommended dose is about 200 gram of dry yeast /1000L of must to start the fermentation. Namely, the starting concentration is about $20 \times 10^9 \times 200/10^6 = 4 \times 10^6$ yeast cells/ml. At this concentration, the added species cells are dominant, and very soon by their multiplication they take over the lead of the fermentation.

Modern winemakers use various species of yeast to best fit the wine type and style at hand. The use of one dominant yeast strain is most beneficial, making the fermentation more consistent, faster and tailored to the winemaker's exact needs.

In contrast to this practice, the "old" tradition prefers to rely on 'wild' yeast found in the vineyard and in the winery equipment. This method lets the fermentation start and proceed to finish 'naturally'. Such fermentation is common mainly in France, as well as in some California wineries. It is believed that by fermenting the wine with various yeast strains, the wine will be more complex, with a higher quality than by only one strain. When using 'natural yeast' fermentation, control over the various parameters mentioned above is not at hand. One of the most abundant strains found in the so called 'natural yeast' technique belongs to another genus, *Kloeckera-apiculata*, which are more active at low alcohol concentrations (about 4%–5% alcohol). In the opposite extreme of tolerance to alcohol, stand the *Saccharomyces-bayanus* species. Some of its strains can tolerate and remain active in up to 16% -17% alcohol.

We find it impractical to detail here the various strains available on the market. Instead, we recommend studying the professional material on this matter, which can be obtained from the major yeast producing companies such as Lalvin, Red Star, Enoferm and others. In these materials one can find all the information needed, regarding the various yeast strains, their origins, microbiological activity, oenological properties, flavor characteristics, and recommended usage.

Killer Yeast

Killer yeast is a mutated strain of yeast, which releases certain toxins into the medium in which it is living. These toxins are lethal to other yeast of the same strain. The killer-yeast is immune to its toxin, while other yeast is either sensitive or unaffected. The phenomena became known in the commercial wine industry in the early 80's. The killer yeast was even developed for commercial use as dry yeast marked as K-1. Now, another groups of killers are known in Saccharomyces genera, named K-2 and K-3. Killers belonging to other genera also have been reported. In respect to 'killing' yeast are grouped as follows:

- **Killers** which are toxic to other yeast
- **Non-killers** which are either sensitive or indifferent to the killer-yeast.

The lethal toxin of the killer is a protein precursor, which when it is absorbed by the sensitive yeast interferes in the proton transport in the membrane cell, making it more permeable to protons. The results are lethal for the host cell.

This matter is of concern in the winemaking industry because in certain cases the results are stuck fermentation. The literature reports on a wide range of killer yeast populations, which cause fermentation to halt. Study on commercially stuck fermentations, showed that in these cases, 90% of the yeast population, were dead, and the 10% viable cells were killer yeasts. This culture of killer yeast was then added into new fermentations and the mortality rate of samples with killer yeast addition was 70% after 3 days, compared to 3% in a control samples. Eventually the killer yeasts take the lead and finish the fermentation, but because of the temporary reduced population, it may take much longer time.

C. Fermentation
1. Yeast Fermentation
The yeast community multiplies itself by budding at a rate controlled by varius media conditions, such as: sugar concentration, temperature, alcohol concentration, nutrients, oxygen and chemicals present. The major visible changes occurring during fermentation are the reduction of sugar concentration, and in parallel, an increase in yeast population. These two changes may serve to sketch the fermentation profile as shown in the following figure:

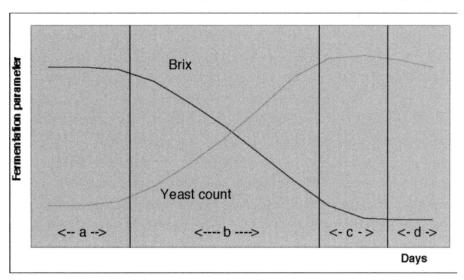

**Schematic profile of the fermentation by two parameters:
Brix reduction and yeast cell count**

After inoculation the fermentation profile is characterized by four phases:

(a) *Lag phase* where the yeast acclimates to the must conditions (high sugar level, low pH, must's temperature, and SO_2 if present).

(b) *Exponential growth phase* where yeast multiply exponentially, reaching maximum population density (about 100 million per cubic centiliter). Sugar concentration in this phase declines fast.

(c) *Stationary phase* where yeast density is in its maximum value, and growth is practically halted. The fermentation rate slows down.

(d) *Decline phase* where nutrients are scarce, the toxic by-products content is high, the viable yeast cells number is gradually decreasing, and the yeast mass settling down.

In order to start a good and healthy fermentation, one has to add enough living yeast cells to the must. By using dry yeast, which is very convenient, the quantity needed is about 20 gram/HL, which is about $10^6 - 10^7$ cells/ml. At maximum activity the yeast cell count reaches up to about 10^8 cells/ml. The population changes during fermentation can be followed by using a microscope with a magnifying power of about 500-1000, allowing good vision on the yeast cells whose size is about 1–2 micron ($1\mu = 10^{-3}$ mm) in diameter.

The yeast should be hydrated first by water/must mixture (10% must) at temperatures of about 38°C–40°C (100°F–105°F), for about 15 minutes. On hydration, the ratio of the dry yeast to water is about 1:10 (1 Kg dry yeast to 10 liters of water/must mixture). Be aware that in 10–20 minutes, the newly hydrated yeast may overflow from its container as a result of very fast fermenting activity. After hydration, it should be introduced directly into the must. Mixing is not necessary.

When using liquid culture instead of dry yeast, one should prepare a starter by heating grape juice diluted by half with water, up to boiling in order to sterilize it. When cooled down to room temperature, add 60 ppm SO_2 and the yeast culture, and mix vigorously in order to aerate it; cover the vessel and mix it once or twice later on. The starter will be ready in about a day or two (at room temperature). The volume ratio of the starter to the must should be about 2% in order to get a good starting fermentation.

Regarding the temperature at the starting stage, there are no problems with red wines, as the temperature is about 22°C–30°C. For white wines, where the fermentation temperature is about 10°C -14°C, it may occasionally take quite a long time before the fermentation starts. It is recommended therefore, to bring the must to a temperature of about 15°C -17°C, then add the yeast, and wait for good signs of fermentation (reduction of one-two Brix unit). At this point, the yeast population has grown enough to carry on the cold fermentation. Then reduce the temperature to the desired range (10°C -12°C). If the yeast in use is highly tolerable to cold fermentation (e.g. Prise de Mousse), the inoculation can be done directly at the cold fermentation temperature.

There are some procedural differences in fermenting white and red musts, because of the skin (cap) management needed in the reds, and the lower temperature needed for the whites. The volume of white must during fermentation should not be more than 95% of the tank volume, so the foam does not overflow when fermentation hits its maximum rate.

In red must fermentation, the volume should not be more than 85%–90% of the tank volume, so as not to overflow when the floating cap is pushed up by the carbon dioxide gas.

The red must should be placed in a fermentation tank equipped with a pumping-over device, which can spray the must taken from the bottom of the tank over the floating cap of skins. This pumping-over is necessary for color and tannin extraction, and should be done three to four times a day during fermentation, and then twice a day until pressing or until the cap settles down (in the case when pressing is delayed for long skin contact). Cap management can also be done mechanically, by punching down the cap in an open tank, but this is an elaborate and difficult procedure.

Other options are now available to treat the cap besides pumping over. One is a closed stainless steel tank that has a device to punch down the cap mechanically. The other option is a 'rotor-fermentor', which is a big horizontal tank that can rotate very slowly around its axis. By doing so, the cap is well mixed with the whole must in regular time intervals. These tanks are quite expensive but they are efficient in extracting the skins, and consequently may save time during the crushing period.

The fermentation rate has to be followed carefully from beginning to end by daily measuring the Brix reading and temperature, and by smelling the tanks. If the fermentation rate does not follow the normal pattern (as shown above in the figure), or if something goes wrong, mainly the fermentation getting stuck, quick action should be taken to prevent serious problems which can develop later on and affect the wine quality.

Also, one of the most common problems during fermentation is what is called 'stinking fermentation'. Experienced winemakers can detect it even at the very early stages, by smelling the fermenting tank. The bad smell is caused by formation of hydrogen-sulfide (H_2S) produced by the yeast in certain circumstances. This phenomenon is quite common and is mainly caused by nitrogen deficiency in the must. When discovered at an early stage of fermentation, the best treatment is to add 100–200 ppm of diammonium-phosphate (DAP) to the fermenting tank. The bad smell may dissipate in one day if discovered and treated early. If it is discovered at the end of fermentation, it is of no use to add DAP, and the best treatment is to aerate the wine by racking. Aerial racking can be done simply, by pumping the wine from that tank to the open top of another one. If the H_2S has not yet been transformed into mercaptan (by interaction with the alcohol), the aeration will be very helpful. When not treated in time, the hydrogen sulfide can become a serious problem that can be treated only by fining with copper. For more details see the General Aspect chapter, section B.

Some wineries with old traditions of winemaking put a piece of copper bar in the must container at the output of the press (in white wine). This brief contact with the copper bar dissolves some copper ions in the must, which then precipitate as copper sulfide when the H_2S is formed, eliminating the stinky odor. This method is quite efficient but has the risk of introducing too much copper into the wine, which may cause copper haze instability later on. Also, copper ions may provoke the yeast to produce more hydrogen-sulfide. So this traditional practice is not recommended. The easiest way to prevent or minimize this problem is first, to avoid any sulfur dust on the grapes by good vineyard management, and second by adding DAP to the must *before* fermentation (at the time of inoculation). The recommended quantity is discussed above.

• • •

A final word on safety during fermentation:

Never add additional must to a fermenting tank from the bottom valves!!!

The result may look like a volcano explosion. The reason is that the fermenting must is supersaturated with CO_2 gas (especially during cold fermentation), and if some perturbation happens in the liquid (like a waves of incoming must from the bottom), it may release the extra dissolved CO_2 all at once, which would look like an explosion. If addition of must is necessary in a fermenting tank, it should be added slowly and from the <u>top</u> of the tank only.

It should also be noted that any stage of fermentation can be checked microscopically, by counting the yeast cells (alive and dead) and by watching the normal budding of the yeast's cells. It is also necessary to follow temperature changes during fermentation and prevent any drastic changes due to failure of the cooling system.

Toward the end of fermentation, analysis of alcohol, residual sugar, volatile acidity, total acidity, pH and sulfur-dioxide, give the winemaker the basic information needed on the processed wine.

2. Stuck Fermentation

Almost any winemaker has had the experience of stuck fermentation at any stage of its progress, from the beginning and up towards the end, when few Brix are still unfermented. Knowing the possible reasons that may limit or inhibit yeast growth at the different phases of the fermentation process, will help to prevent sluggish fermentation from occurring, or in case it happens, to revitalize it. The reasons may be:

Lack of Oxygen : Although fermentation is anaerobic, the use of oxygen at the second phase of fermentation (exponential growth) is of crucial necessity. It is needed for utilizing the steroid *ergosterol* and unsaturated fatty *oleic acid* ($C_{18}H_{34}O_2$), for cell membrane building. Without it the yeast will not be able to multiply and increase their biomass needed to complete the fermentation. In a study where air and nitrogen gas were flushed through experimental samples, the nitrogen-flushed fermentation was extremely sluggish and stuck. On aeration the stuck fermentation was restarted, and kept on until completion. Yeast population ratios between the aerated and nitrogen-flushed samples were over 20-fold higher.

Commercial fresh dry yeast is added at about 20 gr/HL (or about 2Lb to 1000 gallons) of must. At such a concentration it contains a few million cells/ml in the freshly inoculated must.

At maximum activity the yeast cell count, stabilizes at about 100 million/ml, (at the end of phase-II). During the exponential phase-II period the yeast have to multiply 5–6 times (2^5–2^6) in order to increase the population from few million/ml to 100 million/ml. If there is not enough oxygen in the must, the growth will suffer and the fermentation gets stuck. In this case, in order to aerate the must, one can pump about a quarter of its volume over the top of the upper surface of the open tank.

A good practice in white must fermentation is to rack the fermenting must two days after inoculation into another tank. This racking will aerate the must right at the end of the lag period and also homogenize it (yeast culture and nutrient).

Lack of Nutrient: Especially nitrogen in the ammonia form. Some winemakers regularly add DAP (diammonium-phosphate) $(NH_4)_2HPO_4$ to the must before starting fermentation. It is generally less needed in red wine fermentation, but it is recommended in white clear must after lees racking. The added quantity is 10–20 gr/HL of DAP to prevent fermentation problems such as stuck fermentation or 'stinky fermentation' caused by lack of nitrogen. In the case where the fermentation is stuck, it might be useful to add 10-15 gr/HL of DAP and mix it by aerial racking. To complete the nutrient deficiency (rare amino-acids, vitamins and minerals), there are also commercial yeast nutrients, sometimes called 'yeast extract', which contain mixtures of all the necessary ingredients at the right proportions. In cases of stuck fermentation is it advisable to use such extract at quantities of about 20–30 gr/HL.

Temperature: The effect of temperature on fermentation rate is normal as in other biological enzymatic reactions. It increases the rate with temperature in the middle range of activity, and decrease it at the extreme edges, until causing it to become stuck at certain temperature values.

In white wine, which is usually fermented at 8°C –12°C (45°F–54°F), sometimes the lag period may be very long. Acclimation of the yeast culture to low pH, high osmotic pressure of sugar, and very low temperature, might be difficult, causing delay in adaptation. If the yeast is viable, increasing the temperature by a few degrees might accelerate their acclimation and open up the second phase of fermentation. At this stage, fermentation heat is released and as soon as the fermentation is on its way, re-cooling is necessary. To avoid this problem, it is better to inoculate the must when the temperature is higher than the desired fermenting temperature, e.g. to start fermenting at 16°C – 17°C, and only after good signs of fermentation (reduction of 1 – 2 Brix, CO_2 bubbling), to reduce the temperature to the desired one. If the cold fermentation starts well, and then becomes stuck, it works to elevate the temperature a few degrees and add new batch of yeast. After fermentation resumes, the temperature can be lowered again.

At the higher edge of the scale, in fermenting red wine without cooling, or when the cooling system fails, the temperature can rise up to 35°C–40°C (81°F–96°F). Above 35°C the high temperature and alcohol content create a deadly environment for yeast. The yeast has difficulty in maintaining the vital enzymatic reactions, so in red wine fermentation, especially in large tanks, careful watching of the temperature is advised. If the temperature gets too high and the fermentation stops, cooling the must to about 25°C–30°C and adding of a new batch of yeast (after gradual acclimation in the unfinished wine) may restart the fermentation and bring it to completion.

High sugar concentration: When the grapes are very mature and the sugar content is high (above 24 – 25 Brix), the osmotic pressure in the must may inhibit the yeast growth at the beginning of the fermentation. Also, towards the end, when the alcohol concentration gets high, another inhibition factor arises, and the fermentation may become stuck again. To prevent these difficulties in such cases, some measures should be taken right at the beginning. First, select a yeast strain that has high tolerance to high sugar and alcohol concentration. Second, make the environment as convenient as possible for the yeast. Specifically, add DAP and yeast extract, aerate the must well before fermenting, and keep the temperature at a convenient value. If after all these measures are taken, the fermentation still gets stuck, new acclimated yeast should be added. Also, another tolerable strain can be tried.

Fatty acids: It is well known that ethanol is not the only inhibition by-product. It is recognized that fatty acids, mainly *octanoic-acid* $\{CH_3 - (CH_2)_6 - COOH\}$ and *decanoic-acid* $\{CH_3 - (CH_2)_8 - COOH\}$ are responsible for fermentation inhibition. To a lesser extent, their esters are also inhibitors. The medium size fatty acids (6–12 carbons, namely, from *hexanoic* to *dodecanoic* acids) are synthesized during fermentation and are accumulated as toxic agents to yeast growth. They are not by-products of the fermentation per-se, but are "debris" parts, unused during the synthesis of long chain lipid acids, needed for cell membranes. Being hydrophobic, the fatty acids can enter the yeast cell's membrane and interfere with the transport systems between the cell and its medium. The concentration range of octanoic and decanoic acids found in must at the end of fermentation is about 2–10 mg/L. Towards the end, their accumulation can stop the fermentation.

For a long time, it has been known that adding activated carbon can stimulate, and sometimes reactivate, fermentation that has become stuck in its advanced state. In fact, it adsorbs the fatty acids from the must, removing these toxic agents, which then enables the fermentation to proceed.

Dead yeast cells, or 'yeast-ghost' as they are called, can do the same thing very efficiently. The potential of 'yeast-ghost' to restart a stuck fermentation has been verified in experiments. The 'yeast-ghost' can be added prior to fermentation, or after a few days.

Unviable yeast: If fermentation does not start, or continues to be stuck after all the above measures have been tried, a new yeast starter should be introduced. To start such a stuck fermentation, the new starter must be acclimated to the alcohol and other by-products, which are contained in the environment of the stuck must. In such cases the starter should be handled as follows:

After re-hydration with water for ten minutes, add the stuck must to the re-hydrated yeast (about a third of its volume), and measure the Brix of this culture. Allow it to ferment until the Brix decreases about 1–2 units and then add another batch of the stuck must. If it continues to ferment (by Brix measurement), add it to the entire must.

Killer Yeast: Mentioned earlier in this chapter.

3. Fermentation By-Products

Besides ethyl alcohol, which is the major product of alcoholic fermentation, there are many other by-products found in wine as a result of the fermentation process. Some of them are desired and contribute to the wine quality, but some others are harmful and may cause damage to the smell and taste of the wine. We will discuss here the major substances produced during fermentation. Some of them are direct by-products of the fermentation path, while others are formed as a result of other reactions taking place in the must medium during fermentation.

Acids

The following acids exist in wine, although, their origin does not come from grapes. Their accumulating content contributes, together with the grape acids, to the total acidity and the pH in the wine.

Succinic Acid : $COOH - CH_2 - CH_2 - COOH$

This acid is formed during fermentation as direct by-product in all alcoholic fermented beverages and contributes its share to the total acidity. The range of concentration found in wine is 0.5–1.5 gr/L. The acid is very stable and does not change during aging.

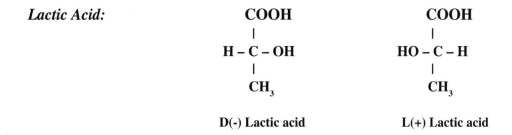

Lactic Acid:

D(-) Lactic acid   L(+) Lactic acid

Lactic acid is formed in wine by two pathways: as a by-product of yeast fermentation, and as the main product in the ML-fermentation of malic acid. D(-) is the absolute S configuration, rotating polarized light to the left. L(+) is the R absolute configuration, with right rotation.

Yeast fermentation produces a racemic lactic acid (a mixture of D and L, called also DL-lactic acid), with some preference to D(-) lactic acid. The concentration range from this pathway is about 0.2–0.4 gr/L. On the other hand, malolactic-bacteria produce *exclusively* L(+) lactic acid from L(-) malic acid, the original malic acid in grape juice. For more details, look ahead in this chapter.

Acetic Acid: CH_3COOH

Acetic acid is the main volatile acid in wine, which may also contain other volatile acids such as formic, propionic and butyric acids (all in very small concentrations). The term VA (volatile acidity) is used to define all these acids as one group in which acetic acid is the dominant one. Acetic acid is formed as by-product of yeast fermentation as a result of the side reaction of acetaldehyde oxidation to ethanol. The normal range of formation is about 200–500 mg/L. At this level it is not noticeable on the palate and has no effect on wine quality. In Botrytised grapes (and wines) the usual concentration is much higher and accepted as part of this style of wine. Some great Sauternes and California late harvest wines contains a typical VA concentrations in the 1.0–2.0 gr/L range.

Acetic acid can also be formed by acetobacter spoilage bacteria, in an aerobic condition. The bacteria oxidize the wine's ethanol to acetic acid at concentrations, which depend on the air exposure and time. In extreme cases it may convert the whole alcohol content into vinegar. Above 1.0 – 1.2 gr/L, VA is noticeable and it depreciates wine quality. The legal limits of VA content in the US, is 1.5 gr/L and 1.7 gr/L for white and red wines respectively.

Alcohols

Methanol: **CH₃OH**

Methanol is not a direct product of fermentation. The source for methanol found in alcoholic fermentation of fruit products, is the fruit pectin, which is hydrolyzed by methylesterase pectin enzymes. Pectin is a co-polymer of galacturonic acid and its methyl ester, through 1 —-> 4 bonding of the acid to the ester molecules. Hydrolysis of the methyl ester yield methanol. Hydrolytic enzymes exist naturally in the must. Addition of pectolytic enzyme to the must in order to facilitate clarification, by breaking the 1 --> 4 bond of the pectin polymer, also increases methanol content. Because pectin is found more in the skin than in the juice, white wines contain much less methanol then red wines, which are fermented with the skin.

A wide survey on reported methanol content in wines around the world showed that the average concentration in white wines is 60 mg/L (in 40–120 mg/L range), and 150 mg/L in red wines (in 120–250 mg/L range). In extreme cases the content may be as high as several hundreds mg/L. The temperature of fermentation does not affect the methanol content in wine, however as said above, pectin treatment and prolonged skin-contact do.

Although methanol is toxic, its concentrations in wines do not show any risk. The fatal toxicity is about 340 mg/Kg body weight. Taking into account methanol content in wine, of about 100 mg/L, and using the toxicity value of 340 mg/Kg, a person whose weight is 140 lb. (~70 Kg), has to drink about 240 liters of wine (!) in order to be intoxicated.

Methanol is metabolized in the body like ethanol (although at a slower rate). The legal limit of methanol in wine in the US is 1000 mg/L.

Higher Alcohols (Fusel oils)

These are the homologous series of saturated alcohols, starting with propanol. Their presence in fermented products has been recognized for a long time, and some mechanisms have been proposed for their production. The major source for higher alcohols are amino acids which by a sequence process of trans-amination, decarboxylation and reduction, are transformed into alcohols as follow:

$$R' - CO - COOH$$

$$R - CH - COOH \longrightarrow R - CO - COOH \xrightarrow{ATP} \xrightarrow{CO_2} R - CH = O \xrightarrow{NADH \quad NAD^+} R - CH_2 - OH$$

$$NH_2$$

$$R' - CH - COOH$$

$$NH_2$$

Amino acid $R - CH (NH_2) - COOH$ is transformed into a-keto acid, which is then decarboxylated to aldehyde (with one carbon less), and then reduced by NADH to its alcohol. For example:

$$(CH_3)_2 - CH - CH_2 - CH(NH_2) - COOH \longrightarrow (CH_3)_2 - CH - CH_2 - CO - COOH \longrightarrow$$

leucine   α-keto-isocaproic acid

$$(CH_3)_2 - CH - CH_2 - CHO \longrightarrow (CH_3)_2 - CH - CH_2 - CH_2OH$$

isovaleraldehyde   isoamyl alcohol

The final result is that an *amino acid* with n-carbons is transformed into an *alcohol* with (n -1)-carbons. When there is a shortage of specific amino acids in the must during fermentation, these trans-amination reactions take place in order to supply the needs of fermentation, and thus higher concentrations of fusel oil are formed as by-products of the amino acids decomposition.

The major high alcohols found in wine are:

n-propanol :	$CH_3 - CH_2 - CH_2OH$
iso propanol:	$(CH_3)_2 - CHOH$
n-butanol :	$CH_3 - CH_2 - CH_2 - CH_2OH$
iso butanol:	$(CH_3)_2 - CH - CH_2OH$
n-amyl alcohol:	$CH_3 - CH_2 - CH_2 - CH_2 - CH_2OH$
3-methylbutanol:	$(CH_3)_2 - CH - CH_2 - CH_2OH$
2-methylbutanol:	$CH_3 - CH_2 - CH (CH_3) - CH_2OH$
n-hexanol :	$CH_3 - CH_2 - CH_2 - CH_2 - CH_2 - CH_2OH$
2-phenylethanol :	$(C_6H_5) - CH_2 - CH_2OH$

68

Their total concentration range in wine is between 100–500 mg/L.

2,3-Butandiol: $CH_3 - CH(OH) - CH(OH) - CH_3$

This alcohol is the major *di-alcohol* found in wine. Its concentration was found to range between 0.3 –1.8 gr/L, with an average value of 0.5–0.8 g/L. Although not yet completely clear, it is a byproduct of fermentation probably from pyruvic acid, or by reduction of acetoin (CH_3-CH(OH)-CO-CH_3).

Glycerol: $CH_2(OH) - CH(OH) - CH_2(OH)$

Glycerol is the major *tri-alcohol* found in wine. It is formed as a byproduct of sugar fermentation, mainly at the beginning of the fermentation, through side-chain reactions, which ends up as glycerol. Glycerol content is related to the amount of ethanol formed, to the temperature of fermentation, and to the yeast strain. It is higher at higher fermentation temperatures, so its content is higher in red wines than in whites. In a study of California wines, an average value of 4.5 gr/L was found in white and rosé wines, and 6.5 gr/L in reds. The content of glycerol reported in a wide variety of world wines, is in the range of 4–12 gr/L. A general conclusion can be drawn that glycerol content in wine is in the range of 6%–12% of the ethanol content. The lower range represents white wines, and the higher range red ones. In late harvest or botrytis wines, the glycerol concentration is significantly higher than in regular table wines. Concentrations of 15–25 gr/L and above have been reported. In this case, glycerol was synthesized by the *Botrytis-cinerea* fungi, and became more concentrated in the dried berries. Some amount of glycerol exists also in healthy grapes (up to 0.5 gr/L). Because glycerol is very viscous and sweet (about 70% of glucose sweetness), it is assumed that it has some contribution to the wine body and mouth feel.

Erythritol: $CH_2(OH) - CH(OH) - CH(OH) - CH_2(OH)$

Erythritol is a *tetra-alcohol* formed from the 4-carbon sugar CHO-CH(OH)-CH(OH)-CH_2OH Erythrose. Note that erythritol is a meso compound (internal symmetry). Its sweetness is about twice that of sucrose. Erythritol is formed during fermentation, and it was found to depend very much on the yeast strain. Its concentration in table wines, which were fermented by Saccharomyces-cerevisiae yeast, is in the 30 -100 mg/L range. Other yeast species such as Kloeckera-apiculata and other 'wild' yeasts can produce several hundred mg/L of erythritol. In Botrytis-cinerea infected wines the concentration of erythritol is in the 50–600 mg/L range.

Arabitol: $CH_2(OH) – CH(OH) – CH(OH) – CH(OH) - CH_2(OH)$

Arabitol is a *penta-alcohol*, which was reported to be found in wine in the range of 10–60 mg/L in white wines and 30 -110 mg/L in reds. In high quality German wines infected by Botrytis cinerea the concentrations found were much higher, in the hundreds of mg/L range, and in several cases up to 2300 mg/L of arabitol. It is assumed that arabitol is produced by yeast other than Saccharomyces-cerevisiae, which are grown with the noble rot infected grapes.

Mannitol: $CH_2(OH) – CH(OH) – CH(OH) – CH(OH) – CH(OH) - CH_2(OH)$

Mannitol is an *hexa-alcohol* formed from fructose during fermentation and is found in wines in the range of 100–400 mg/L. Late harvest German wines showed very high concentrations of mannitol from hundreds of mg/L up to several grams/L. In this case it is related to the delaying of the harvest. (Mannitol can also be formed by lactic acid bacteria, which is always followed by acetic acid formation. In such cases the wine is considered spoiled by the acetic nose).

Sorbitol is another hexa-alcohol, an isomer of mannitol. It's found in wines in a concentration range of 50–200 mg/L, and in late harvest German wines, as much as a few hundred mg/L up to 1000 mg/L. As with other polyalcohols it is associated with the noble rot grapes and wines.

Aldehydes and Ketones

Aldehydes and ketones are formed during alcoholic fermentation. The main representatives, which are found in wine, are *acetaldehyde*, *acetoin* and *di acetyl*.

Acetaldehyde: CH_3-CHO

Acetaldehyde is the last major component in the chain fermentation process from sugar to acetaldehyde, which normally is then enzymatically reduced to the final product, ethanol. But some of it may not be reduced, and remains in the wine in a concentration range of 20–200 mg/L, with an average content in table wines of about 70 mg/L. In aged wines its concentration may increase by a slow non-enzymatic oxidation of ethanol to acetaldehyde:

$$CH_3\text{-}CH_2\text{-}OH \xrightarrow{\text{(o)}} CH_3\text{-}CHO$$

At bottling, when wine is highly exposed to air, some extra amount of acetaldehyde may be formed, which gives the wine some flat taste. This phenomenon is sometimes called "bottle-sickness", and it disappears with time. Enough free SO_2 before bottling will minimize this effect.

Flor-sherry yeast under aerobic conditions oxidizes ethanol to acetaldehyde (and some other aldehydes such as propionaldehyde, isobutyraldehyde and isovaleraldehyde). These aldehydes, including acetaldehyde, are not by-products of fermentation and have no connection to fermentation at all. In *sherry wines* the average concentration of acetaldehyde is very much above 200 mg/L. Acetaldehyde has a very distinct nose, which characterizes *sherry wines*, but it is not thought to contribute to the quality of *table wines* when it is noticeable by its smell.

Acetoin: $\quad CH_3 - CH(OH) - CO - CH_3$

This ketone is formed as a by-product of the fermentation. During fermentation the acetoin concentration reaches a maximum value of about 100 mg/L, which later decreases (probably by reduction to 2,3-butanediol), to about 5–20 mg/L. In dessert port-type wines where the fermentation is halted in the middle by alcohol addition, the acetoin concentration is found to be higher than average, in the 20–50 mg/L range.

Diacetyl: $\quad CH_3 - CO - CO - CH_3$

This di-ketone is also a by-product of fermentation. It is found in very low concentrations. Red table wines, which had undergone malolactic-fermentation, had an average of 2.8 mg/L, and those which had not, had a lower average range of 1.3 mg/L. Thus, malolactic-fermentation contributes *additional* diacetyl to the regular post-fermentation content. Significant detection thresholds for the taste of diacetyl were shown to be about 1 mg/L. It is estimated that at 2–4 mg/L range, diacetyl contributes to the quality of red table wine by its buttery aroma, making it more complex. Above this range the quality is reduced, and is recognized as an off-flavor nose.

Esters

Esters are the product of the reaction between alcohols and acids. In wine they are classified into two categories: *neutral* and *acid* esters. *Neutral-esters* are formed through an enzymatic process. They include acetate, butyrate, oxanoate and other esters. *Acid-esters* are formed by a regular chemical esterification between alcohols and wine acids at low pH.

In wine, esters come from three sources:

(1) Originally produced by the grapes, in very small amounts.
(2) By-products from the fermentation process (neutral-esters).
(3) Production from the aging of the wine; a very slow chemical esteri fication (acid-esters).

The neutral-esters are considered to be volatile aroma components, also called *volatile esters*. The acid-esters, mainly tartaric and malic esters, are called *non-volatile esters*.

Our interest here is in the second group, which are fermentation by-products, and are important factors for wine nose and quality. The concentrations of the volatile esters in wine are low and their organoleptic perception is described as a fruity-type aroma, except for *ethyl acetate,* which has the highest concentration in wines, and whose contribution to wine quality is considered negative.

Ethyl acetate: $(CH_3 - CH_2 - O - CO - CH_3)$

It is always accompanied by acetic acid, and in fact it contributes the distinctive vinegar acescence to wines when its concentration is higher than 160–180 mg/L, in comparison to acetic acid, which starts to be noticeable (only by taste as a hard sourness) at above 750 mg/L.

The actual concentration of ethyl acetate found in wines as a by-product of fermentation is about 30–70 mg/L. On the other hand, spoiled wines by *acetic acid bacteria* at aerobic conditions have much higher concentrations of ethyl acetate where the main product of this spoilage, is acetic acid.

Higher volatile esters

Beside the above negative ester, young wines contain higher volatile esters, which have great significance mainly in young white wines in relation to their fresh fruity aroma.

Also they are very important components in brandy, where their concentrations are raised in the distilled brandy. When young wine is aged, its volatile esters are gradually decreased by hydrolysis, loosing its fresh fruity aroma. Some of the most dominant esters are:

Isoamyl Acetate: $(CH_3)_2$-CH-CH$_2$-CH$_2$-O-CO-CH$_3$ at concentration range of (2 – 6) mg/L, with an odor threshold of (0.16–0.23) mg/L.

Ethyl Octanoate: CH_3-$(CH_2)_6$-CO- O-CH$_2$-CH$_3$ at concentration range of (0.5 – 2) mg/L, with an odor threshold of (0.24–0.58) mg/L.

Ethyl Hexanoate: CH_3-$(CH_2)_4$-CO- O-CH$_2$-CH$_3$ at concentration range of (0.2–1.5) mg/L, with an odor threshold of 0.076 mg/L.

Ethyl Decanoate: CH_3-$(CH_2)_8$-CO- O-CH$_2$-CH$_3$ at concentration range of (0.1 – 1) mg/L, with an odor threshold of (0.5 – 1) mg/L.

This is only a partial list of about 300 esters, which have been found and listed in wine.

4. Residual sugar

Classification

The term *residual sugar* refers to any significant concentration of sugar that is contained in wines. It may vary from 2–4 gr/L in very dry table wine, up to 50–150 gr/L in port style wines or late harvest botrytised dessert wines. At any fermentation, when going smoothly there is some residual sugar left over. This residual sugar is mainly unfermentable pentose sugars, namely, *arabinose* and *rhamnose*. There may also be some unfermented fructose and glucose. If the fermentation is stopped before all fermentable hexose sugars have been consumed, more fructose will be found than glucose, because the glucose rate of fermentation is slightly higher than that of fructose. Also, besides sugar, there are other sweet components in wine, such as glycerol, at a concentration range of 4 -12 gr/L with about 70% of the sweetness of glucose. Ethyl alcohol has also some slight sweet taste, so even when the fermentation has come to "complete" dryness, some people experience a sweet taste in dry table wine, especially whites, but also in reds.

In good quality containing-sugar wines, no addition of can-sugar is accepted. The sweetness should come only from the natural grape sugar, either by sugar left over after fermentation, or by addition of grape juice or grape concentrate to the dry wine to make it sweet. This is of common practice; it is legal and well recognized.

There are many types of wines containing sugar in the market. They can be classified into four major categories:

• **Off-dry table wines** with about 10–30 gr/L sugar. In certain table wines, especially if they have fruity and fresh character (e.g. White Riesling, Chenin-Blanc, French-Colombard) it is best balanced with some sweetness. In Muscat related varieties (e.g. Muscat of Alexandria, Muscat Canelli, Gewurztraminer, Sylvaner, Symphony), it is almost essential to leave some residual sugar in order to match with the very spicy and powerful perfumed aroma of these varieties. In certain cases it is necessary to mask the bitter after-taste that sometimes characterizes these varieties. Also, residual sweetness is common in the 'blush' wines ("blanc de noir").

• **Semi-sweet wines** mainly whites with some higher sugar content of about 30 – 50 gr/L The low-medium sugar content is necessary to well balance the high acidity content of these wines. It is typical to grapes grown in cool climates.

• **Sweet dessert wines** such as: Port, Sherry (Oloroso), Madeira, Marsala, Vermouth and others. Each wine in this category has its special character, depending on its unique version of production. The residual sugar content in these wines is usually between 50–100gr/L.

- **Late harvest wines** with two main styles:
  1. ***The German late harvest style,*** (Auslese, Beerenauslese, Eiswein) which are made of over-ripe grapes where the sugar level gets higher than the regular level. In certain cases where the grapes become shriveled and almost dry, the sugar level is very high. The most popular varieties in this version are White Riesling and Gewurztraminer. Hungarian Tokay can also be classified here.
  2. ***The Sauterne (Bordeaux)*** *style* wines, which are made of late harvest grapes infected by *Botrytis Cinerea* mold ("Noble Rot"). The infection, which usually causes the grapes to become 'rotten' and undesirable as table wine, may in certain weather conditions (high humidity at infection time, and dry and warm in the following days) develop in such a way that they cause the berries to lose their water without rotting. The sugar level (and all dry extract) becomes very high to about 30–50 Brix. The Botrytis not only concentrates the dry extract in the berries but also gives some very special aroma characteristics to that style of wines. The wine made from these grapes ends up with about 50 -150 gr/L residual sugar. The most common varieties for this type of wines are Sauvignon Blanc and Semillon, but White Riesling or Gewurztraminer is also suitable. When such wine has to be made, its processing is quite different from table wine. The right time to pick the grapes is most crucial. If picked too early, the sugar will not be concentrated enough, and if picked too late, the grapes might be totally rotten. Several pickings may be necessary according to the bunch "Noble Rot" development, and even individual berry picking is practiced in some wineries.

     The pressing of the shriveled berries (whole clusters, no destemming/ crushing), needs high pressure for a very long time, and many cycles of pressing. The juice contains a very high percentage of solid lees (which will be difficult to clear up), and a high content of aldehydes, volatile acids, and glycerol. No settling is done and the general practice is to add about 100 ppm SO_2 before fermentation, which will probably be fixed by the aldehydes. Diammonium-phosphate (DAP) is also recommended (200 ppm), to ease the fermentation. The inoculation is best done with yeast culture rather than with dry yeast, in order to acclimate the culture gradually to the very high sugar level of the must. It is also preferable to start the fermentation at a moderate temperature (18°C–20°C) and only after it has developed well, reduce the temperature to 10°C–14°C. After fermentation has stopped at a certain sugar level and the wine has been racked off, fined and stabilized, it can be aged in oak barrels for 6-18 months before bottling. In certain places the must itself is fermented in oak barrels.

Addition of SO_2 is necessary to maintain about 20–30 ppm of free SO_2 during all the processing stages up to bottling. Because of the initial high SO_2, the total SO_2 in late harvest wines is usually considerably higher than in regular table wines. 200-250 ppm is a common level in late harvest wines.

It must be mentioned that the grapes in the German late harvest version may sometimes also be infected with Botrytis.

Stopping Fermentation and Preservation

There are several techniques to stop the fermentation before dryness and to preserve it from re-fermenting. Stopping fermentation is the 'natural' method to leave a certain amount of residual sugar in the finished wine. It can be done in any of the categories, fully sweet, semi-sweet dessert wines, or off-dry table wines.

Deep Cooling: Most suitable for premium table wines. If this technique has been chosen, right from the beginning the winemaker should make the yeast's condition not very comfortable. Under some stress the yeast will be weak enough to be easily controlled. No nutrition should be added before fermentation (lees settling is obvious and using bentonite prior to fermentation is also possible). Yeast inoculation can be done with half the regular amount, and low fermentation temperature 10°C -12°C is the main regulation factor.

Although any yeast strain can be used, the yeast strain that is easiest to stop with cold-shock is preferred, e.g. *Epernacy-2*. This strain is also good for fruity and highly aromatic white wines. Under all these conditions, the fermentation may last longer than usual, 4 to 6 weeks. Toward the end of the fermentation, the alcohol and sugar analysis should be made and, at the desired sugar concentration, the temperature should be drastically lowered under 4°C (39°F). The fermentation will then terminate. If the fermentation is stopped while the yeast is very active and the rate of the fermentation is still high – there may be some delay before the final termination. If a strong must chiller is not available to immediately cool the must, it will take some time for the temperature to drop down just by reducing the temperature control. In such a case, deciding when to stop the fermentation should be based on extrapolating the Brix reading above the desired residual sugar level. For such extrapolation, a daily record of the Brix level is necessary.

In about a week or two, the wine has to be racked off from the yeast lees, sulfited, bentonited, and cold-stabilized. After cold stabilization, the wine should be filtered, and aged in a tank until bottling, at low temperatures not exceeding 5°C–8°C.

Natural Stopping: In some cases, just by imposing stress on the yeast, the fermentation may stop by itself. The stopping is unpredicted with no control on the desired residual sugar level in the wine. In late harvest must, the conditions are such that the fermentation is always under stress. The nutrition is usually too low, and the high sugar concentration is an inhibition factor for the fermentation, especially when the alcohol be-comes high enough, depending on the residual sugar concentration. So, in most cases, the high Brix late harvest must will stop its fermentation naturally.

Alcohol Fortification: Another technique to stop fermentation and leave residual sugar, is by addition of alcohol to the fermenting must up to final concentration of about 18% (v/v), where the yeast cannot tolerate anymore and dies. The addition of alcohol is done at the desired sugar concentra-tion, usually at 5%–10% in dessert wine production, like Port, Sherry, Madeira and others. After the addition of alcohol, the wine is well mixed, racked from the yeast lees and cold-stabilized. The high alcohol and sugar concentration will preserve the wine from re-fermenting during the rest of the wine processing and in the bottle.

Sweetening Dry Wine: There is a different technique to provide residual sugar in wine, which is less troublesome, and less caution is needed. This is most suitable for regular, less expensive, off-dry table wines. In this technique the wine is fermented to dryness, racked from the yeast, sulfited and bentonited. After bentonite racking and filtering, a calculated amount of sugar is added as grape juice concentrate, to bring the wine to the desired sugar level. Stabilizing and aging is carried on as any regular wine, but until bottling the wine should be kept at low temperature of less than 5°C–8°C to prevent re-fermentation. Another way to sweeten dry wine is by adding 'sweet reserve' (unfermented grape juice) which has been kept in the cellar since harvest. The 'sweet reserve' is preserved by lees settling the clear must, sulfited up to 400–500 ppm of SO_2, and stored at cool temperature at 0°C–3°C.

Less care is needed if the sugar addition is made just before bottling, where the addition is followed immediately by micro-filtration. The only risk with such late sugar addition is that it may change slightly the wine's stability in the bottle. However, if the wine is to be consumed young and early, this risk is minimal.

Sugar Preservation in the Bottle

Any fermentable sugar is a major microbiological instability factor in bottled wine and special attention has to be paid to maintain it. This section deals with some ways to preserve wine from re-fermentation:

Yeast Inhibitors: The most common agent to inhibit fermentation is *sorbic acid* or its salts. Before bottling, about 200–250 ppm of potassium-sorbate can be added. The sorbate has an inhibitory effect on yeast growth, and should be used in no less than 200 ppm in order to be effective. In complementary action with the sorbate, 30–40 ppm of free sulfur-dioxide is most recommended. The sorbate has almost no offensive taste or odor at these levels, especially in sweet wines. Its taste threshold is around 150 ppm, and its limit in the United States is 1,000 ppm. The main objection to the use of sorbate is the 'geranium like' odor which may develop by lactic bacteria from sorbic acid to produce *2-ethoxyhexa-3,5-diene*. To some people this compound may be pleasant, to others unpleasant. The SO_2 presence with sorbate is a good assurance that ML-bacteria will not be active to carry on that reaction. The sorbate is widely used in sweet wines, which have not been fortified or pasteurized.

Pasteurization: A very old and efficient technique to prevent re-fermentation of sweet or semi-sweet wines. It is not recommended for quality table wine because the heat will destroy its freshness and fruity aroma. It also may contribute a caramel taste to the wine. The lowest damage is caused by flash pasteurization at about 80°C for a few seconds and immediately cooling. After the pasteurization, the wine may become cloudy because of protein precipitation, so fining with bentonite and filtration is necessary. However, by doing so, the wine may become un-sterile again from the tanks and the equipment used, so another pre-bottling pasteurization is needed. To avoid it, the wine can be bottled directly after the bentonite racking through pad filtering and micro-filtering to the bottling line.

Sterile Filtration: This technique is based on mechanical removing of the yeast from the wine, and keeping it under conditions that make fermentation very difficult to restart (low nutrient, alcohol, SO_2, and almost null yeast population). Up to bottling, when sterile filtering is done, the wine has to be checked regularly for signs of re-fermentation (CO_2 bubbles, microscopic test). At bottling, the sterile filtration (0.45μ) prevents any wine spoilage microorganism from getting through. In the modern technology of winemaking, this is the most used and effective technique to prevent re-fermentation in the bottle.

D. Malo-Lactic Fermentation

1. General Aspects

Malolactic fermentation is caused by a malolactic bacteria, which belongs to three genus: *Lactobacillus, Pediococcus*, and *Leuconostoc*. Although, all these genera may be found in wine, the first two genuses are undesirable. They can cause off-flavor and unpleasant after-taste. Also they produce high concentration of acetic acid. The species of interest (the 'good guy') is *Leuconostoc oenos*. Inoculated malolactic fermentation had become a common practice in the modern technology of winemaking, although it has been known to naturally occur since Pasteur's times in the middle of the nineteen-century. Inoculation technology has started to develop commercially in the last two-three decades, and now it is done on almost any red premium wine, and certain white wines. The ML commercial material is sold as a starter-culture or as lyophilized freeze-dried bacteria ready for use right after re-hydration.

Chemistry of ML-fermentation

Malolactic bacteria live on glucose and other sugars and organic acids, but they also carry on the transformation of malic acid to lactic acid:

$$HOOC - CH_2 - CH(OH) - COOH \longrightarrow CH_3 - CH(OH) - COOH + CO_2$$

through three possible pathways:

(1) $HOOC-CH_2-CH(OH)-COOH \longrightarrow HOOC-CH_2-CO-COOH \xrightarrow{-CO_2}$
      Malic acid   oxaloacetic acid
    $CH_3-CO-COOH \longrightarrow CH_3 - CH(OH) - COOH$
     pyruvic acid   lactic acid

(2) $HOOC-CH_2-CH(OH)-COOH \xrightarrow{-CO_2} H_3-CO-COOH \longrightarrow$
      Malic acid   pyruvic acid
    $CH_3-CH(OH) - COOH$
      lactic acid

(3) $HOOC-CH_2-CH(OH)-COOH \xrightarrow{-CO_2} CH_3-CH(OH) - COOH$
     Malic acid   lactic acid

In this transformation L(-) malic acid is converted into L(+) lactic acid. If malic acid had been added to must for acid correction {the commercial malic acid is a racemate, a mixture of L(-) and D(+)}, only the L(-) malic acid will be consumed by the ML bacteria, leaving the D(+) malic acid untouched.

In this transformation the 4-carbon di-carboxylic acid is decarboxylated into a 3-carbon mono-carboxylic acid, and CO_2. The term 'fermentation' used in this microbial activity refers to the CO_2 formation with no real justification, because no real energy is produced in this process.

Thermodynamic calculations show that the free energy difference (ΔG) between products and substrates is practically negligible (-2 Kcal/mole), which brings up the question: What benefit does the bacteria gain by malolactic transformation? Later study suggests some slight excess of the oxidized NAD^+ coenzyme, which is an important factor in oxidation-reduction metabolism processes, which may enhance the bacteria's initial growth rate in the presence of malic acid.

Another option is that some extra ATP is formed and expressed as an increase in D-lactic acid formation from glucose by malolactic bacteria in the presence of malic acid. On the other hand, it may be that the only benefit for the bacteria is the reduction in acidity and increased pH caused by the malic to lactic transformation. The convincing answer is still a mystery.

The reduction in titratable acidity by malolactic fermentation (MLF) is in the range of 1–3 gr/L, which may be quite important in softening high acid wines. MLF also adds some metabolic products, which may improve wine quality. Diacetyl {CH_3-CO-CO-CH_3} is such major by-product. A survey on Australian and Portugal red wines revealed an average content of about 3 mg/L of diacetyl in wines which had undergone MLF, compared to an average of about 1.5 mg/L in those which were not.

Induction of MLF can be spontaneous or by inoculation. Spontaneous MLF can happen at any time when the right conditions are met (from fermentation and up to the bottle).

There is a very simple analytical method to determine whether MLF has occurred. This test is a qualitative determination only, but for all practical purposes it is good enough.

Benefits of MLF

What are the benefits of having malolactic fermentation occur in wine production?

Reduction of acidity: when it is too high especially in cool regions. MLF can reduce the titratable acidity by 1–3 gr/L, and increase the pH by 0.1–0.2 units.

Flavor changes: are expected to result from ML fermentation, as some composition changes occur during this process. Beside the main product lactic acid, other compounds such as diacetyl, acetoin, 2,3-butandiol and volatile esters are also produced.

Studies on the question regarding the contribution of MLF to wine quality brought about controversial results. In many cases there was no significant difference between wines with and without MLF. The reduction in acidity is well recognized, and the 'buttery' flavor of the extra diacetyl presence in MLF wines was also noted. But the contribution to the overall quality, and the recognizable difference between MLF wines and those without it, remains for additional studies.

Microbial Stability: in regard to malolactic infection, stability is gained by having it occur during wine processing, and not in the bottle. This advantage is significant when the wine conditions are favorable for MLF to occur, namely, high pH, low SO_2 level, and also in coarse filtered wines (appropriate for red wines). Other microbial infections are more likely to occur with the increased pH level caused by the MLF.

Aside from these benefits, MLF is not desirable in fruity, fresh style wines (white or red), as the MLF might change the wines' fruity character. Also it is not desirable in low acid wines, which after MLF will be more flat and unbalanced.

2. Factors Affecting Malo-Lactic Fermentation

Several factors affect the malolactic bacteria growth:

Temperature: Below 15°C (59°F) the growth rate is practically zero. Warmer conditions are needed to stimulate growth. The range of 20°C (68°F) to 25°C (77°F) is optimal.

pH: Malolactic bacteria are sensitive to low pH. At pH = 3.0–3.3 *Leuconostoc oenos* can live but would do better at higher pH. For other genera (*Lactobacillus, Pediococcus*) it is almost impossible to grow at pH below 3.2. At pH above 3.6 they grow well, spoiling the wine by producing large quantity of diacetyl, acetic acid and mousy taint. As a result of MLF, the TA is reduced and the pH increases (an average reduction of 0.15% TA and rise of about 0.1–0.2 pH units).

SO_2: Sulfur dioxide is a very effective inhibition agent against malolactic growth, even in its bound form. At pH = 3.5, presence of 30 ppm free SO_2 was found to completely inhibit ML growth and activity. The same concentration of acetaldehyde-bisulfite (bound SO_2) was also very effective although its action was delayed. The inhibition is effective even at 10 ppm of free or bound SO_2. In other studies the inhibition appears less effective at low concentrations, with concentrations above 50 ppm SO_2 needed to be effective.

Nutrient: Nutrient deficiency can inhibit ML growth. Early racking after yeast fermentation (if MLF was not carried on with yeast fermentation simultaneously), may leave the wine low in vital nutrients, which can prevent the bacteria from growing.

Inhibitors: (a) *Fumaric acid* {COOH- CH=CH- COOH} is used as an inhibitor for ML bacterial growth when it is undesired. The mechanism of action is not yet clear, but it is very effective when used as an additive in concentration of 600–700 mg/L (at pH = 3.4 where these experiments were done). It is evident that pH and fumaric acid have a combined effect and the concentration of fumaric acid which is needed to inhibit ML fermentation, depends on the wine pH. In another study the use of 500 to 1500 ppm of fumaric acid at three pHs showed the combined inhibition effect of these two factors. Use of 500 ppm which looks quite effective at pH = 3.5, is certainly not enough at pH = 4.1. These measurements were done in wines kept in favorable conditions for MLF (except fumaric acid), namely, without SO_2, at 25°C, and no removal of yeast lees after fermentation. In real wine conditions, inhibition would be more effective. So, in clear, low pH white wine (3.0–3.4), with 20–30 ppm free SO_2 the 500–700 ppm of fumaric acid would be quite effective. If the pH is higher, higher concentrations of fumaric acid are needed. Addition should not be done prior to yeast fermentation because fumaric acid will be degraded by the yeast. So, if MLF is undesired, and use of fumaric acid is chosen to do the job, it is recommended to add it to the young wine after first racking. Fumaric acid is difficult to dissolve in wine, which may cause some practical difficulties. Its solubility in water at 25°C, 40°C, and 100°C, are 6, 10 and 98 gr/L, respectively. The high temperature solubility is recommended for use when needed. Fumaric acid showed no negative effect on the wine sensory score when added up to 1.5 gr/L range.

(b) *Fatty acids* are known to be microbial toxic factors. Their inhibition effect on malolactic fermentation in wine range concentrations showed definite effect, especially a mixture of hexanoic, octanoic and decanoic acids. As individual acids their inhibition was much less effective. Dodecanoic acid (C_{12}) was also found to be an effective ML inhibitor.

3. Inoculation of Malo-Lactic Fermentation

In cases where it is not left spontaneously to occur, MLF can be initiated simultaneously with yeast fermentation, or after it was over. Post fermentation MLF can be more difficult for several reasons: low nutrient content, high alcohol, SO_2 added after fermentation, and cold temperatures. Simultaneous MLF and yeast fermentation has become very common in recent years. The bacteria have access to fresh must nutrients, with low alcohol at the start. This enables the wine to finish the MLF more or less at the time of alcoholic fermentation with less risk of other microbial spoilage (SO_2 can be added right after racking). When MLF is carried out simultaneously, SO_2 should be avoided during fermentation, or added in less then 25 ppm. *It should be noted that there is a possibility of high volatile acidity in simultaneous fermentation (especially when yeast fermentation is somehow delayed) because of the sugar consumption by the ML bacteria, which produces acetic acid.* D(-) lactic acid is also produced by the ML bacteria from sugar consumption.

On inoculation with healthy culture of dry bacteria (10^8–10^{10} cells/gr), it may happen that the viability falls down substantially, by the suddenly 'hostile' wine conditions. It is therefore recommended to acclimate the culture with increased portions of the must/wine before inoculation is made to the bulk. A recommended procedure is as follows:

1. Apply 10 grams of dry bacteria or culture in 100 ml of apple juice (pH = 5.5 and 25°C).
2. After a few days transfer to 500 ml of apple juice (pH = 4.5 ; same temperature).
3. After the same period as above, transfer to 10L of diluted grape juice/water 50%. In all three stages don't add any SO_2. You can check the viability of the culture by paper chromatography (disappearance of malic acid).
4. When the primary culture is ready, increase the inoculation volume (with a fermenting wine if MLF is done during alcoholic fermentation, or with a still wine if done afterwards). Firstly, add to a 200L at its natural pH and temperature above 16°C. Then increase the inoculation at increments of 2–5% culture in the new inoculated wine. Each scaling up takes a few days before the next one.

In recent years, new dry yeast cultures of pure *Leuconostoc-oenos* species were developed which have the advantage of being ready for use without any acclimation stages. These freeze-dried cultures can be added after re-hydration directly into the wine, saving the whole troublesome inoculation procedure. Of course, some minimum precautions should be taken, such as minimum pH value (above 3.1), minimum temperature (above 16°C), and maximum SO_2 concentration (below 40 ppm total). Each commercial culture has its specific inoculation instructions.

When MLF is *not desired*, measures have to be taken to prevent it. Otherwise, it will probably happen. In the winery, the best measure is low temperatures (below 10°C), and enough SO_2 level above 20 ppm free). MLF in the bottle is a very serious problem. It will spoil the wine by CO_2 bubbles, sediments, and probably bad smell. Before bottling, enough SO_2 should be added, and sterile filtering at 0.45μ should be employed to assure that the malolactic bacteria will not start fermentation in the bottle.

Following-up ML Fermentation: The test is performed by paper chromatography with a special eluting solvent, which is made from a mixture of: n-butanol (100 ml), water (100 ml), formic acid (10.7 ml), and indicator solution (15 ml). The indicator solution is 1% of *bromocresol green* in ethanol. The mixture is shaken and placed in a separatory funnel. The aqueous phase (lower) is discarded. A drop of the tested wine is spotted on the paper by capillary tube. Thick and dense chromatography paper is advised. When the paper is placed in a closed container with some small volume of the eluting solvent (about 1-2 cm height), all the spotted wine acids are moving upwards on the paper by the eluting solution. When the climbing liquid level reaches the top of the paper (few cm below it), the paper can be removed from the container and allowed to dry. After a few hours, a chromatogram is developed on the paper, showing the major acids present in the wine.

A reference chromatogram of the major wine acids (tartaric, malic, lactic) can be prepared and run on the paper with the wine samples, to indicate each acid position in the chromatogram. The chromatogram will indicate whether malic acid is still present in the wine or not.

CHAPTER IV

CELLAR OPERATIONS

Chapter IV: Cellar Operations

A. Racking

Wine is racked off a couple of times during processing. Racking is another word for decanting, meaning transfer of liquid above its solid sediments. In wine processing, solids are called *lees* and may contain yeast cells, pulp, tartarate crystals, fining particles, proteins and tannins. In the racking operation, two major changes occur: *clarification* and *aeration*.

In general, the main principle in racking is simple: *as little as possible*. This means that the racking should be done only when certain operations are necessary such as: leaving sediments at the bottom of the tank, or when blending wines, or any wine transfer that is needed during processing.

Wine tanks contain a few outlets with valves. The bottom one can rack off all the liquid from the tank. The second valve is located about 20–30 cm above the bottom one. In most cases the lees layer is somewhere below the level of the second valve. Therefore, racking off the wine (or white juice), from its solids content is done from this valve. When air starts to be seen at the hose, the racking is over, and the pump should be shut off. The rest of the wine may contain low or very high quantity of solids. In the last case, either the bottom is drained off, or some of the wine can be saved. The best way to save this wine, is to let it out from the bottom valve to a small container and let it to rest for a few days before racking it again above its solid layer. When the sediment layer is low and dense, in order to save extra work, the hose can be connected to the bottom valve and carefully racked off, to the rest of the wine. Some portion of the sediments will be transferred into the racked wine. Some tanks have the option of an inner bent tube placed through the second valve, inside the tank that can be rotated from the outside by the operator, to alter the level of suction. By gradually lowering the inner tube down it is possible to pump out the wine exactly to the lees surface level. In this way, the racking is continued until this point, and the rest is drained off. The lees layer is watched from the tank's top opening. Two people are needed for this operation. This devise is very helpful and most recommended. The other side of the hose is connected to the receiving tank, either at the bottom valve if no aeration is desired, or at the top of the tank when aeration is desired. All connections between tanks, hoses and the pump should be sealed well, with no leakage of liquid out, or leaking in of air.

Racking operation can also be used to administer and mix some of the necessary materials that are used during wine processing such as fining agents, sulfur dioxide, acids addition, or blending with another wine. When this is the case, after finishing the racking, in order to complete the mixing, the wine can be circulated in the tank from bottom to top.

Any aeration during racking introduces oxygen into the wine at concentrations from tenths of mg/L in careful racking, up to 5–6 mg/L of oxygen in over-aerating racking. For white wine, less oxygen is better, to prevent browning and preserve freshness. In red wine, certain amounts of oxygen are necessary during the process of aging, so racking with some aeration is a good practice. In order to minimize oxygen absorption during racking, the receiving tank can be filled prior to racking with CO_2 gas through the bottom valve, for a couple of minutes. The CO_2, which is heavier than air, will stay on the wine surface and will protect the incoming wine when it fills the tank. Nitrogen is another good option. The major rackings are:

Zero racking: in white wines. We call it zero racking because it is done before fermentation. This is the clear juice racking from the must lees, which is necessary for good white wine making.

First Racking: after fermentation has terminated. This racking roughly clarifies the new wine from the very dense sediments of mostly yeast cells and pulp. It should be done shortly after fermentation has stopped, to prevent extraction of dead yeast constituents into the wine, like protein metabolites, amino acids and hydrogen sulfide. In red wine, there is a large quantity of seeds in the lees and if left un-racked for a long time, the wine extracts much of the tannins. In certain styles of wines, whites or reds, the first racking is delayed for a long time, regardless of the above considerations. In white wines it is done especially in selected lots of Chardonnay (or Sauvignon-Blanc in certain wineries in Boudreaux), which are barrel fermented and left over after the fermentation for a couple of months on the lees ('sur lees'). In red wine, some vineries keep the yeast lees for several weeks before racking.

Second Racking: in white and red wines, the second racking is needed to further clear the wines from the sediments left over from the first racking.

Third Racking: in white wines, after fining (with bentonite) and cold-stabilizing (can be done in a row, and then racked off at ones). In this case the lees consists mostly of fining particles and tartarate crystals. This is also a good chance to use this racking for blending. In reds, the wine is racked to barrels, to start its barrel aging.

Fourth Racking: in whites, this racking is for the first filtering. Because filtering is in fact transferring wine from one tank to another one through a filter, we consider filtering as racking. In certain cases filtering can be done right after bentonite and cold stabilizing, namely, in the third racking. By doing so, this racking can be avoided. The racked wine is then stored in tanks until bottling. In reds, during the barrel aging of red wine, racking from barrels to other barrels is done in intervals of four to six months. In this racking the barrel's bottom sediments, which contain yeast cells, fine pulp particles, bi-tartarate crystals, and some polyphenols-protein colloids, are cleared off. The empty barrels (after racking off) are washed with water and refilled from other racked barrels. In many cases it is done differently, namely, by racking the barrels into a tank, cleaning the barrels, and then racking the wine back into the barrels.

Fives Racking: in white wines, from the tanks to the bottling line through sterile filters. In reds, from the barrels back to the tanks, to unify the barrel's wine and to blend it for further storing and aging before bottling.

Sixth Racking: red wines are racked from the tanks to bottling through filters.

This racking scheme represents a general framework. In some cases, because of certain reasons, some lots must be treated and racked more than the above description.

Because the wine is exposed to oxygen in each racking, in order to protect it from being oxidized, addition of SO_2 during each racking is advised. The actual quantity can be determined by measuring the free sulfur-dioxide before racking, then adding enough to have about $20 - 25$ ppm of free SO_2 after racking. For details look in General Aspects chapter, section A. These small additions will accumulate to the total SO_2 in the wine.

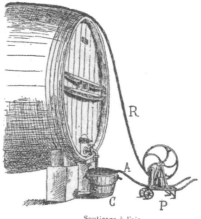

Soutirage à l'air.

B. Stabilization

In the wider aspect, this term refers to operations that prevent cloudiness and settling of particles in the bottle. The causes of cloudiness and solid particles in the bottle can be: protein coagulation, polyphenolic colloids, protein-metal haze (iron and copper) and tartarate crystallization. Protein and polyphenolic stabilization will be discussed in the next 'Fining' section. Iron and copper haze, which is very rare now, as all cellar equipment is made of stainless steel, will be discussed in the General Aspect chapter, section B. Here we shall deal with tartarate stabilization.

1. Tartaric acid dissociation

Tartaric acid (represented here by H_2T) is the major and unique acid in grapes and wines. It dissociates into bitartarate (HT $^-$) and tartarate (T $^=$) ions, and forms salts with the major cation in wine–potassium, to form potassium-bitartarate (KHT):

$$H_2T \xrightarrow{\quad K_{d_1} \quad} HT^- + H^+$$

$$HT^- \xrightarrow{\quad K_{d_2} \quad} T^= + H^+$$

The K_d values and their corresponding pK_a (at 25°C) are:

$$K_{d1} = 9.1 \times 10^{-4} \quad ; \quad pK_a' = 3.04$$
$$K_{d2} = 4.25 \times 10^{-5} \quad ; \quad pK_a'' = 4.34$$

Based on these pK_a's and using the formulation for week acids (see Pre-harvest chapter section A) the distribution of tartaric acid components (HT, HT⁻, T⁼), in water solution, at wine pH range can be calculated, and is shown in the following figure:

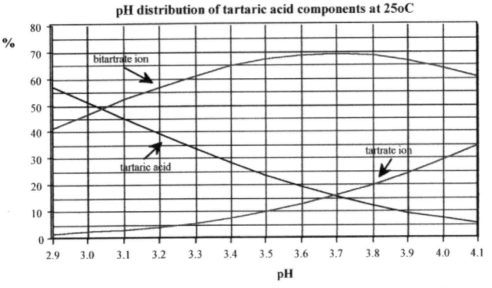

(Note the maximum concentration of bitartarate ion at pH = 3.7 which will be discussed later).

2. Potassium bitartrate

The potassium bitartrate salt, COOH-CH(OH)-C(HOH)-COOK, called also *cream-de-tartar,* forms crystals which belong to the orthorhombic space group. Their specific weight is very close to 2 gr/cc. The crystals are soluble in grape juice, but are less soluble in the alcoholic solution formed after fermentation takes place.

Any ionic substance will precipitate when the *concentration-product* of that salt exceeds its *solubility-product* at the specific conditions under study. *Concentration-product* (CP) is the product of the actual ions concentration of the salt in solution at a given condition.

Solubility-product (SP) is the product of the ions concentration in solution when the solute is in its maximum possible concentration, namely, in its saturation concentration. In formal presentation, a salt in solution dissociates to its ions:

$$AB <===> A^+ + B^- \quad ; \quad \text{then } CP = [A^+] \times [B^-]$$

Where: $[A^+]$ and $[B^-]$ are the A^+ and B^- molar concentrations of the ions respectively.

SP is equal to CP when the salt is saturate in that solution. There are three possible relations between CP and SP:

(1) When CP < SP: the salt concentration is below its saturation point. Then, the salt in solution is stable and will stay so, as long as the conditions are unchanged.

(2) When CP = SP: the salt in solution is in its saturation point, a state where the solution can not contain any more of the solute in solution, unless the condition are changed so it will be possible.

(3) When CP > SP: the salt in solution is over-saturate, an unstable solution, which a minor change in the solution conditions might cause the solute to precipitate.

Let us check the case of potassium bitartarate in wine. The concentration-product of potassium bitatrate in wine is defined as:

$$CP = [K^+] \times [HT^-] = [K^+] \times [\text{total } H_2T] \times (\% \text{ of } [HT^-])$$ (in the wine's pH and alcohol concentration).

The wine pH, its alcohol concentration, and the temperature are the most important factors affecting the solubility of potassium bitartarate in wine:

pH: We showed above the various distribution components of tartaric acid vs. the pH values in wine pH range. As can be seen in the figure, as the wine pH gets higher, the tartaric acid portion gets lower, and the % of bitartarate ion gets greater, which makes the concentration-product of potassium bitartarate to become larger too. For example, at pH = 3.1, bitartarate ion is 52% of the total tartaric acid present, while at pH = 3.4 it is 65%. Hence, at higher pH, the concentration-product of tartarate is higher too. This means that at the same molar concentration of potassium bitartarate in two wines, the one with the lower pH may be stable, while the one at higher pH might be unstable and precipitate.

Alcohol: Potassium bitartarate is quite stable in grape juice, although its CP is close to its SP (below of course). During wine fermentation, it become unstable and starts to precipitate in the presence of alcohol in the wine. The reason for the precipitation is an effect known as salting-out. In this effect, the less-dielectric solvent (alcohol) added to water, reduces the solubility-product of ionic substances, causing them to precipitate if their concentration-product were close to their solubility-product. In this case the SP of potassium bitartarate became smaller in the new medium—wine, and from the previous state, namely, CP < SP, a new state CP > SP was established, causing part of the salt to precipitate, until equilibrium CP = SP is reached.

Temperature: Also influences very much the solubility-product of any salt including potassium bitartarate. The solubility-product at high temperatures is higher then at lower temperature. This means that the salt can change its status from CP < SP to CP > SP with just a reduction of the temperature of the solution.

So, all these factors, namely, the pH, the alcohol content, and the temperature affect the solubility and therefore the wine stability towards tartarate precipi-tation. In real wine, other factors also influence the solubility of potassium bitartrate, such as other organic acids, and polyphenols in red wines, which inhibit precipitation. (This is why in old red wines, which were stabilized in the winery before bottling, there are always crystals of potassium bitartrate. As the phenolic compounds polymerize and their content is reduced after long aging, their inhibition of potassium bitartrate crystallization diminishes, and fine crystals of the salt are seen on the cork and in the bottle).

3. Cold stabilization
The concentration ranges of tartaric acid and of potassium ions in ripened grapes are about 2–6 gr/L and 1–2 gr/L respectively. The concentration of

potassium bitartarate in grape juice is close to its saturation, and as the fermentation proceeds (forming more alcohol), the salt becomes supersaturated. In such a case, precipitation is inevitable.

Deposition of potassium bitartarate in the bottle is considered to be an aesthetic defect and measures are taken during the winemaking process to eliminate this instability potential. The most common method for tartarate stabilization is chilling the wine to such a low temperature, that its concentration-product (of the salt left after precipitation) will be much less then the solubility-product at normal environment temperatures. The wine will therefore reach stability. When a cold stabilization takes place, changes in wine acidity and pH occur as well. These changes are caused by the change in the concentration of tartaric acid and potassium ion. Always the acidity is lowered because tartaric acid is removed from solution. But the pH changes depend on the initial pH, where the key value is pH = 3.7. Observing the above figure of tartaric acid dissociation vs. the pH, shows that at this pH the bitartarate concentration is at its maximum. At initial pH values *lower* then 3.7 any reduction in the bitartarate concentration brings back the equilibrium curve of bitartarate by lowering the pH. For example, if the initial pH was 3.4 and the reduction in bitartarate concentration by the chilling was from 65% to 60%, then the bitartarate equilibrium line at 60%, is at pH = 3.3, which is lower than the initial one by 0.1 pH unit.

On the other hand, if the initial pH is *higher* then 3.7, any reduction in the bitartarate concentration brings back to the equilibrium curve by increasing the pH. For example, if the initial pH was 3.8 and the reduction in bitartarate concentration by the chilling was from 69% to 65%, then the bitartarate equilibrium line at 65%, is at higher pH = 3.95.

As a result, when stabilization occurs, the pH changes according to the initial pH value, namely, below pH = 3.7 the pH will decrease, and above pH = 3.7 it will increase. But, in the pH range around 3.7, where the curve is almost flat from say pH =3.55 to pH = 3.75, there will be practically no change in pH by the chilling.

There are some formulas and tests, which can be done to determine the optimal temperature and duration of wine chilling. The time needed for achieving stability depends on the kinetics of the bitartarate crystallization under the cold stabilization conditions, namely, the temperature, the initial concentration-product of potassium bitartarate (which depends on the potassium-ion concentration, tartaric acid concentration, the pH, and the alcohol content), and on the nucleation factors which may increase or decrease the crystallization rate. In general, crystal growth rate in solutions is controlled by two parameters: the transportation of the substance from the solution to the growing crystal (diffusion), and by the rate the solute ions migrate on the crystal surface to their final position in the crystal. Diffusion rate is mainly determined by the temperature, while the second parameter depend on the crystal particles' size, and impurities in the solution (which may enhance or block the crystallization process). These "impurities" are wine colloids such as proteins, fining agents, polyphenols, pectin and other hydrocarbons. They may interact with the K^+ and the HT^- ions, blocking their way on the crystal surface.

On the other hand, the impurity particles may serve as nucleation sites for the supersaturated solution to crystallize. It is therefore not accurate to predict the stability of wine just from its concentration-product, as could be done if the medium were just an aqueous-ethanol solution. We think this is of no importance, and that simply by keeping the wine at temperature range below 0°C and above the freezing point (-5°C), for two–three weeks, the potassium bitartarate stabilization will be set. The cooling can be done either by passing the wine through a heat exchanger (chiller), resulting in quick cooling to the desired range, or by lowering the temperature of the tank by its temperature control, which takes a couple of days. In our opinion, the latter option is preferred, especially in white wine, because it saves the need for one more wine transfer. To decrease super-saturation, and to ease the crystallization and precipitation, it is recommended to combine the bentonite treatment (see later) with the stabilization operation. In such a combination, the bentonite is added with the right amount of sulfur dioxide (25–35 ppm), mixed well, and then after a day, the temperature control can gradually reduce the temperature. The fining particles will facilitate the settling. Cold stabilization can be hastened by seeding with crystals of potassium bitartarate. Most of the potassium bitartarate precipitates on the sidewalls, and the bottom of the tank.

After two weeks, the wine can be racked and cold filtered before bottling. The winemaker saves one racking, in comparison to when these two operations (fining and stabilization) are done separately. It is possible to check the total acidity (and, better still, the tartaric acid level, if possible) before stabilizing and during cooling. By following these changes, one can know when settling has come to equilibrium. Also, the bitartarate crystallization rate can be monitored, by measuring the electrical conductivity of the wine. It will reduce as long as the bitartarate concentration is reduced, making it possible to determine the end of crystallization.

4. Calcium tartarate

Calcium tartarate (CaT) is less soluble in wine then potassium bitartarate. And because natural calcium concentration in wines ranges between 30–200 mg/L, it may contribute to the wine tartarate instability. The sources of calcium in wines are from natural content in grape juice, and from additions to the wine by various agents containing calcium, such as $CaCO_3$ (for acid correction), bentonite, filter-aid, or by concrete tanks storage. When the concentration-product of calcium tartarate in the wine is greater then its solubility-product, it will precipitate, although, its precipitation rate is very slow and may take months to achieve stability. Also, racemic calcium tartarate (with DL-tartaric acid) is much less soluble (about 35 mg/L in water at 20°C), than the L-calcium tartarate (about 300 mg/L at 20°C). The *natural grape juice L-tartaric acid* undergoes slowly some racemization with time, and therefore, a calcium tartarate, considered stable, might become unstable after certain time. This is because its concentration-product exceeds its new solubility-product, which is about 8 times lower. In white wine, if consumed young, within 1 to 2 years, there should not be a problem. In red wine the tartarate is more stable in solution because of the tannins and, in any case, some sediment in red wines after many years in the bottle is tolerable and accepted. This fact can be used in stabilizing wine with regard to calcium tartarate. Addition of racemic DL-tartaric acid will cause faster precipitation. The practical addition dose is about three times the calcium concentration (at the range of 100 to 500 mg/L of DL-tartaric acid).

5. Inhibition and promoting bitartarate crystallization

The function of *inhibition* agents in this case is to prevent the treated wine from undergoing bitartarate precipitation, and therefore, becoming stable in this respect. The principle of their action is to interact with the crystal surface in a way that prevents any further growth. By doing so, although the wine is supersaturated, it will remain stable. Cold tests done on treated wines with inhibition agents do not show any tartarate precipitation, and therefore they are considered to be stable.

In contrast to inhibition agents, *promoting* agents aim to increase the crystallization rate of bitartarate at room temperature, and thereby increase the wine's stability. The advantages of using such agents (inhibitors and promoters) are that stabilization can be achieved immediately with no cold stabilization operation (saving energy cost), and also it can assure stability of the wine for a certain period of time.

Metatartaric acid: Addition of very small quantities of metatartaric-acid (about 100 mg/L) inhibits crystal formation in wines for a certain time. The inhibition is a result of interaction of this substance with the growing crystal, preventing any bitartarate ions from joining the surface. After addition to wine, the metatartaric acid is slowly hydrolyzed back to tartaric acid, losing its inhibition activity. The efficiency and duration of inhibition depends on the storage temperature. The cooler the wine, the longer precipitation can be inhibited. At room temperature it may last up to one year. The use of this agent to stabilize wine from bitartarate precipitation is therefore good for short-aging wines such as whites, rosé, and light reds. Metatartaric acid is made from tartaric acid by heating it above its melting point at 170°C to form a polymer of tartaric acid. After cooling, the solid product is ground to form a fine powder. Before adding it to the wine (prior to bottling), it should be dissolved in cold water (about 200 gr/L).

Carboxymethylcellulose (CMC): Another colloidal substance that interacts with tartarate crystals and inhibits their growth very efficiently is the polymer *carboxymethylcellulose*:

This substance is a cellulose chain, in which its OHs are partially substituted by sodium carboxymethyl groups. The length of the polymer chain (n = 25–250) and number of sodium carboxymethyl groups per sugar monomer (up to three such substitutions per each sugar monomer), determine the solubility of the substance in water. The commercial substance has good solubility in cold and hot water, and forms a very viscous solution at quite low concentrations. It is approved as a food additive and it is used mainly in the liqueurs production to make it thicker and smoother (at concentrations of about 1000–2000 mg/L).

In wine, it acts like metatartaric acid in stabilizing wine from bitartarate precipitation, but it has the advantage that it is very stable and does not lose its efficiency with time and higher temperatures. It was found to be effective at very low concentration ranges (25–50 mg/L). Wines treated by this agent show very good stability in the cold test.

Calcium seeding: This agent contains calcium tartarate and calcium ion in a powder form. When added to wine, the calcium ion reacts with the excess tartarate ion (T $^=$) to form insoluble calcium tartarate crystals. The calcium tartarate particles in the reagent, function as seeding nuclei for crystal formation. The concentration range for using this agent (commercially known as "Koldone") is between 0.5–2.5 gr/L. The exact quantity can be found in a laboratory test by increasing concentration of the agent in samples of the wine. The agent is mixed well with the wine samples (at room temperature) and left over for 24 hours. Then the wine samples are filtered and placed in cold temperature (about -3°C to -8°C) for about 10 hours (freezing may occur). After a day at room temperature the samples are checked for crystal formation. The sample, in which crystals were not formed, with the minimal amount of agent added, is the one to use for stabilizing. For convenience, the agent is used as a 12% slurry in water. After preparation, the slurry should be left over-night and remixed just before adding it to the wine while mixing it well with a mixer or by pumping. The wine should be then left to settle down, and a stability test should be done. If the wine is stable it can be racked and filtered.

The last thing to remember is that after blending wines, there might be instability potential and a cold stability test is recommended.

C. Fining

Fining is the operation of adding certain substances into wine, in order to improve its appearance (turbidity, color), to remove some off-flavor, and to prevent some potential instability which may develop in the future by some wine components. The fining agent reacts with the specific wine component either chemically or physically, to form a new component that tends to separate from the wine either by precipitation or filtration. Because the interaction is not always of a pure chemical nature, the quantity ratio between the wine components and the fining agents is not strictly constant, and a preliminary lab test is necessary in each case prior to the agent addition. In certain cases the fining agents may themselves contribute either an off-flavor or add instability to the wine, or remove and reduce some of its natural flavor. So, the use of fining agents has to be done carefully and with the minimum necessary doses, in order to avoid over-fining.

1. Protein stabilization

The protein content of grape juice varies in the range of 100–800 mg/L. During fermentation the protein content may either increase or decrease (up to about 40%).

The amino acids in proteins act as amphoteric dipolar ions. They can be represented by the formulation $NH_3^+ - CH(R) - COO^-$ called *zwitterions* (double ion). The dipolar ion behavior of the protein is a result of the association/dissociation reactions of its amino acids in the acidic/basic medium:

$$NH_3^+ - CH(R) - COOH \xleftarrow[OH^-]{H^+} NH_2 - CH(R) - COOH \text{ (Amino-acid)}$$

$$NH_2 - CH(R) - COOH \text{ (Amino-acid)} \longleftrightarrow NH_2 - CH(R) - COO^-$$

At wine pH (acidic) the protein's amino acids are therefore positively charged (upper part of the equation). The solubility of proteins in an aqueous medium depends on the pH of the solution and the *isoelectric point* (PI) of the protein. The isoelectric point is the pH at which the positive and the negative forms of the protein are at equal concentrations. At this specific pH the protein is electrically neutral, and therefore it is *least soluble* at its isoelectric point. Wine pH range (3.0 to 4.0) is very close (slightly below) to the isoelectric points of most of its proteins. Also, the alcohol in wine reduces the proteins solubility. These two factors may cause some protein instability in wine and it means that precipitation of the proteins is only a matter of time, mainly at higher temperatures. The result is a colloidal cloudiness and haze. Sauvignon Blanc wines for example, are known to have high concentrations of unstable proteins. The protein colloidal particles that are positively charged may interact (under certain conditions) with negatively charged particles to form neutral particles that can coagulate (flocculate) and settle down. This is the principle of protein removal from wine by fining. The two major agents used for that purpose are *bentonite* and *silica-gel*.

***Bentonite*:** Bentonite is a mineral material composed of aluminum-silicate anions $(Al_2O_3 \cdot SiO_2)(H_2O)_n$ neutralized by cations such as calcium, sodium, potassium and magnesium. Sodium bentonite is the most common and effective, but calcium bentonite is also used. The microscopic structure of the mineral, reveals very small plates of the order of 0.1μ with a very high surface area (about 500–1000 square meter per gram of bentonite). The plates can adsorb water and swell to form a colloidal suspension. Being negatively charged, bentonite particles can interact with positively charged protein molecules on its plate's surface, and consequently the protein is removed from the liquid by precipitation of the bentonite particles. The interaction of bentonite with proteins is very fast, in a matter of minutes. Afterwards, the adsorbed proteins slowly settle by gravitation of the bentonite particles to form bentonite lees at the bottom of the tank. The mechanism of adsorption between the protein and bentonite is assumed to be of a fixed number of absorption sites on the solid bentonite plates, which bind to the protein molecules on a one to one basis. This leads to describing the bentonite-protein interaction by the Langmuir equation of adsorption, as was shown in a model wine solution. This is the common case of protein removal. But, in those cases where the protein segments have isoelectric points which are very close to the actual wine pH, their interaction with bentonite is minimal, and therefore such proteins are difficult to remove by bentonite. Unfortunately, these proteins are also the most unstable in solution, and are the most likely to cause protein haze. Also, there are protein fractions, which have their isoelectric points *below* the actual wine pH, and therefore they are *negatively charged* and consequently they will *not* interact with bentonite at all. Fortunately the number of such fractions is very small. Because of these considerations, it is worthwhile to keep in mind as a general guide, that the lower the wine pH, the higher is the efficiency of protein removal by bentonite. Or in other words, it should be understood that the use of bentonite to remove proteins from wine has some limitations. And in cases where high quantities of bentonite are used in order to achieve stability, the wine might be partially stripped of its flavor components, gaining an earthy taste, with an over-fined wine as a result.

When the fining is done on grape juice, the reduction in proteins and amino acids may cause fermentation problems (lower fermentation rate, stuck fermentation, high residual sugar). It may also affect the hydrogen-sulfide formation. Pre-fermentation fining with bentonite is done for proteins removal from *white wines* solely. It does not make sense to do so in red wines, as will be discussed in the next paragraph (on phenolics fining).

Preparation of bentonite for use is done by hydration of its particles in hot water for at least several hours (preferably one day) before adding it to the wine. The recommended concentration is 5% weight / volume. The hydrated bentonite swells and forms a very viscous slurry gel. The usual concentration range of bentonite usage in wine is 0.2–1.5 gr/L.

The amount of bentonite used should be minimal to fulfill its duty. An excess of bentonite reduces color and aroma, and may leave some off-flavor. It may also increase the pH. One should test in the laboratory for the right amount of bentonite to use by testing a series of wine samples (say 0.5L each), with increased quantities of bentonite: 0.2, 0.4, 0.6, 0.8, and 1.0 gr/liter (equal to 0.4, 0.8, 1.2, 1.6, 2.0 liter of slurry/HL of wine). All samples (with an untreated one as a control) should be well mixed with the bentonite slurry and allowed to stand overnight. After settling, the samples should be decanted, filtered (1 micron filter), and checked for protein stability by heating the samples for 6 hours at 70°C to 85°C. Following this, the samples are left to cool at room temperature, and are checked after three days with low intensity light for the cloudiness of each sample. The appropriate amount of bentonite will be the clear sample with the least bentonite addition.

When added to the wine, bentonite acts best at moderate temperature (15°C to 25°C). The sediment volume of bentonite is about 2%–3% of the wine volume, which may cause some wine loss. To minimize some of the loss, it is recommended therefore, to treat first with bentonite and then to follow with the cold stabilization treatment. The advised procedure is: bring the wine temperature to 15°C–25°C, and add bentonite while racking the second lees operation. Allow the sediments to settle down for a few days, and then do the cold stabilization and filter it (cold). The seeds of potassium tartarate, which are formed during cold stabilization, will help to reduce the bentonite lees level to a minimum.

Some varieties such as Sauvignon Blanc or Gewurztraminer will need more bentonite in order to be protein stabilized. Excess use of bentonite (over 0.8gr/L) may strip some of the aroma.

In certain cases, a compromise will have to be made between prospective protein instability and reduction of the wine's aroma.

A stability test is needed after bentonite treatment to assure wine stability. The test is done by heating the wine to certain temperatures for a certain length of time. The unstable protein will form a noticeable haze. Many reports in the literature deal with the recommended temperature and time needed for this test. From our experience and the literature, the wine should be heated to 70°C–80°C for 6 hours, then cooled back to room temperature, and compared with a reference wine that had not been heated. If haze is seen in the heated wine, it is still not stable, and more bentonite is needed.

Silica-gel: Silica fining agent is an aqueous colloidal suspension of silicon-oxide prepared from sodium silicate ($Na_2SiO_3.H_2O$) by addition of strong acid to form silicic acid. The amorphous silicate particles (typical diameter of 50A°) are held in suspension by the negative charges on its surface. The charge is the result of the dissociation of the surface hydroxyls but it is also stabilized by sodium ions added to it:

$$-\underset{\underset{/}{|}}{Si}-O-\underset{\underset{/}{|}}{Si}\begin{array}{c}\overset{O^-}{} \\ \overset{OH}{} \end{array}\ldots$$

The commercial silica agent (called *kieselsol, baykisol, klebsol*), is a milky-like suspension of 30% (by weight) of SiO_2. The negatively charged particles can interact with positively charged protein molecules, the same as bentonite does, or as tannin can do. The combined action of silica and gelatin in clarifying cloudy wines is very efficient and common.

Silica is a very clean fining agent in the sense that it can be made at high purity and it does not contain other components, which might affect the wine aroma and flavor. On using it for white wine clarification and stabilization, addition levels of silica are used in the range of 0.1–0.25 ml/L (when using the commercial 30% liquid form). When added with gelatin to white wines, the recommended level range of the gelatin addition is 20–50 mg/L. In such case, the two agents should be added to the wine separately, the silica being the first. Coagulation and separation from the wine begins in a very short time after addition. If bentonite is also used, it is better to add it first, and then the silica followed by the gelatin. Excellent clarity can be achieved with silica-gelatin combination in difficult cloudy cases such as botrytized wines. The use of silica for clarification is also common in other beverages. Silica agent was approved for use in foodstuffs in the U.S. by the FDA, and by the International Oenological Codex.

The main advantage of using silica for fining is as a replacement of tannin in the gelatin fining of white wine. The tannin is added in order to avoid excess gelatin content in the fined wine, but it may also add some astringency to it. The silica is neutral with no taste effect at all, so over-fining is not a risk.

2. Tannin reduction

The presence of tannins in red wines at high concentrations leads to astringency and a hard mouth feeling. In white wines sometimes they may cause bitterness and a yellow-brown color. It is therefore desirable in certain cases to reduce their content in red or white wines. The principal method to do so is based on the interaction between tannins and proteins or protein-like molecules.

The practical application of tannin-protein interaction has been in use for a very long time in the leather industry where hide reacts with tannin to produce the so-called 'leather'. The reaction between tannins and proteins is a hydrogen-bond interaction, between the phenolic hydroxyl as a proton donor and the peptide's keto-imide carbonyl as an acceptor:

$$- \overset{\text{II}}{\underset{O--H--O}{C}} - NH -$$

It has to be emphasized that the protein-tannin bonds are not ionic, nor covalent bonds, but rather of a hydrogen-bond nature. The hydrogen bonding is multivalent, due to the many sites of hydroxyls and carbonyls in the interacting molecules. Cross-linked complexes, which are insoluble in water, may be formed. The tannin-protein complex may be soluble or insoluble in water, depending on its concentration and on the molar ratio between the two. When tannin is in excess, the complex will probably precipitate, but when protein is in excess, the complex might remain in solution.

The nature of the complex, namely its bond strength, solubility, reversibility, and microbial resistance, depends on the protein (amino acid composition, its molecular weight), the tannin (polymeric size), and the reaction conditions (pH, ethanol, ionic strength). The main phenolic target for removal from wine, are the procyanidins (polymers of flavanol-3 and flavone-diol 3,4) in the molecular weight range of 500 – 3000 atomic units, which are mainly responsible for astringency in red wine.

Gelatin: Gelatin is a protein product made from animal tissues (bones and skin). Collagen is the raw material, which after processing (partial hydrolysis, evaporation, drying and grinding), is sold as a white powder or in sheets. The molecular weight of commercial gelatin for wine fining is in the range of 15,000–150,000 and it contains about 85% proteins and up to 2% mineral residue. The rest is water. The gelling power of gelatin (namely, the resistance to deformation of the gel made under specific conditions) is expressed in *bloom number*, which ranges in wine gelatin from 80–150 (the recommended range is 80–100). Gelatin can absorb water up to ten times its weight. The isoelectric-point of gelatin is in the range of 4.8–5.2, above wine pH, and therefore it is positively charged. Gelatin interacts mainly with *polymeric phenols* rather than with low molecular weight phenolic compounds (namely, with greater number of hydroxylic sites per molecule in the poly-phenols), and therefore its fining effect depends on the wine age and oxidation state. It is more effective on *aged red* wine than on younger ones. When applied, it can reduce astringency, some bitterness, and polymeric anthocyanins in red wines. It is very useful in reducing the harshness of pressed wine. Some care is needed so as to not strip the wine flavor and character by over-fining. The range of gelatin addition to red wines is 50–100 mg/L. Up to 200 mg/L can be added in certain cases to pressed wine.

The dose range of gelatin in fining white wines is 20–50 mg/L. It can clear cloudy wines that are difficult to be cleared with bentonite. It is also effective in reducing bitterness after-taste in white wines. In order to prevent protein over-fine (and instability), silica-sol (or tannin) is added in combination with gelatin at a quantity range of 10–25 ml of kieselsol 30% commercial solution in a hectoliter of white wine. The addition of kieselsol has to be done before the gelatin fining (a day). As with other fining agents, laboratory tests must be carried out in order to determine the right quantity before it is applied. Also the actual quantities of gelatin/kieselsol have to be determined by testing. Racking and filtration of the fined wine has to be made after one-two weeks. By such fining, excellent clarification can be achieved. There is no legal limit for gelatin addition in wine.

Gelatin solution is made by adding 1% gelatin powder or sheets, to almost boiling water with constant mixing on addition. The gelatin is a viscous gel solution, which should be added to the wine while it is still hot (it will solidify on cooling). On slow addition to wine, thorough mixing is essential to ensure even contact with the entire wine volume. As in red wine, racking and filtration should be done within one-two weeks.

Isinglass: Isinglass is a protein extract made from the swim bladder of certain fish. The commercial product looks like transparent chips. The molecular weight is about 140,000 and its isoelectric point is in the range 5.5–5.8, well above wine pH. It is used in white wine clarification at very low concentrations (10–50 mg/L) with excellent results. There is no need for tannin or silica-sol addition in combination with isinglass, as with gelatin (see above). The material interacts mainly with the *monomeric phenols* rather then polymeric ones. There is no legal limit for isinglass addition to wine.

The preparation of an isinglass solution is as follows: To 10 liters of cold water add 10 grams of tartaric acid and 4.4 grams of potassium metabisulfite (250 ppm SO_2) and mix until all dissolved. While mixing, add slowly 100 grams of isinglass. Stir again a few more times in the following day, where upon the solution becomes viscous and gelly-like. The solution can stand and be in good condition for days. It contains 10 gr/L of isinglass. With such solution, each cc contains 10 mg of the agent, and for example, if an addition of 20 mg/L is needed to 4000L of wine, 8L of the isinglass solution is required. Racking and filtration can be done in a week or two.

Casein: Casein is a milky protein, which has been used as a wine-fining agent for very long time. It is prepared from milk as powdered potassium caseinate, which is water-soluble. In low pH (as in wine) it is not soluble and will precipitate as soon as it is added. It is used in white wines to reduce phenolic bitterness, over-oaked white wine, some off-flavors and to lighten brown color in oxidized wine and pinking. The doses needed are in the range of 50–250 mg/L of potassium caseinate. Milk can also be used for fining, especially low-fat milk to avoid the milk-fat introduction into the wine. About 2–10 ml/L of milk is the concentration range of milk usage.

To prepare the agent for use, make a 2% (20 g/L) solution of potassium caseinate in warm water and stir patiently for a long time. Leave overnight and stir again until it becomes completely dissolved. The solution is good for a day or two. Each cc of it contains 20 mg of the agent. When added to wine it should be well mixed. Before racking, bentonite can be added and then racking and filtration can follow.

Egg-White (albumin): The substance is the white section of eggs, which contains about 10% protein (albumin and globulin). This fining agent is solely used for red wine fining, and is considered as the "best" for softening and polishing red wines. The protein interacts mainly with *higher polymeric* phenols rather than lower ones or monomers. The usual doses are one–two eggs per barrel of red wine (225L). To prepare the egg white from fresh eggs, break the eggs and separate the whites from the yolks. Collect all the white and mix them in 0.5%–0.9% salted water (table salt). The salt is needed to dissolve the globulin of the egg white, which is soluble only in salted water. Albumin is soluble in pure water. The eggs/water proportion is about 15 eggs per 1L of water. Mix well with a mixer and use it the same day. When the solution is added to the wine (in barrels) it has to be stirred very well. Racking should be done in about a week. Dried powder of egg white is available also, but it is less favored. About 3 grams of dry powder is equivalent to one egg.

PVPP (polyvinylpolypyrrolidone): This is a synthetic polymer whose structure is:

PVPP is a higher molecular weight polymer of PVP (polyvinylpyrrolidone). The exact molecular weight is difficult to measure because the substance is not soluble in any common solvent. The commercial substance is sold in granular form with mean particle size of about 100μ. Being polyamide, PVPP interacts with phenolic compounds as other agents do. But contrary to most of them, it reacts specifically with *low polyphenols* such as monomers and dimers (e.g. cathechin, anthocyanin). The interaction takes place between the PVPP carbonyl group and the phenolic hydroxyl. Because the PVPP is insoluble, the phenolic molecule adsorbs on its surface and precipitates out of the solution. In white wine, PVPP is useful for reducing brown color and pinking, and as a preventative measure when such risk is forecast. It will contribute to color stability in sensitive blush wines. When used for color reduction in white wine, a combination with carbon is more effective in many cases. It helps also to settle down the carbon particles. In many cases PVPP can also reduce certain off-flavors and bitterness. Performing a trial to see if it works in a specific case might be beneficial. Dose range in white wine is 100–700 mg/L. In red wines it is less used but it can reduce bitterness, and brighten the color. The dose range in red wines is 100–200 mg/L. Addition of PVPP can be done at any stage of production, from must to pre-bottling. PVPP is a gentle fining agent, which almost does not strip wine aroma as other agents do. PVPP is added directly into the tank while stirring. Mixing with wine or water (5%–10% slurry) prior to addition to the tank is also helpful. It interacts very fast and settles within a few days. The legal limit of PVPP in the US is 800 mg/L.

3. Non-specific fining

Carbon: Activated carbon is made of small particles, which have an extremely high surface area, which ranges about 500–1000 square meter/gram. The carbon surface is capable of adsorbing both in the gas and liquid phase, by physical adsorption, which is caused by Van Der-Waals attraction forces. The characteristics of the absorption, namely the concentration of the adsorbed substance per weight unit of carbon, (as a function of the adsorbed substance concentration in the medium under study), can be described by either Langmuir or Freundlich equations. The formulae work in either medium, liquid or gas phase.

In wine, carbon can be used for removing off-flavors and odors of various kinds, and to decrease browning or pinking in white wines. The carbon type most suitable for decolorizing is marked as *KBB*, and the deodorizing type is marked as *AAA*. As mentioned above, it works well in combination with PVPP in both tasks. The adsorption on the carbon surface is very fast, so the results are felt immediately. Due to its absorption capability, carbon will also absorb aroma and flavor components, reducing their concentration in the wine, so care should be taken when carbon is used, and moreover, it should be used only if no other means to solve the problems is possible (e.g. $CuSO_4$ can reduce the hydrogen sulfide or mercaptan very successfully instead of carbon).

The range of usage is between 1 to 5 gr/HL (10 – 50mg/L) for color treatment and 5 to 25 gr/HL (50–250 mg/L) for off-odors removal. The exact quantity required should be determined in the laboratory prior to its addition to the wine. In the laboratory test, the samples should be mixed with carbon, filtered after an hour, and checked for improvements of color or smell. When added to the bulk wine, at temperature ranges of 15°C–25°C, the black carbon powder is administered directly and well mixed. It will settle in a few days. The use of carbon was not restricted in the US until recently. It is now limited to 3.0 gr/L which is by far, above the practical range.

A summary of the various fining agents, their use, dose range, and application methods, is given in the following table:

Agent	Use for	Dose range	Preparation and use
Bentonite	# clarification of white wine. # protein reduction in white wine. # stabilization of white wine against protein haze formation.	0.1 - 1.5 gr/L Less bentonite is needed in low pH wines.	Add slowly 50 gr/L of bentonite (5%) to almost boiling water while stirring. Let stay for a day with occasional mixing. It will become a viscous slurry solution. The solution is good for a few days. 1 cc of the slurry contain 50 mg of bentonite. Do not fine very cold wine (below 15°C).
Gelatin	# clarification of white wine. # tannin & bitterness reduction in white wine. # tannin & astringency reduction in red wine. # tannin & astringency reduction in pressed red wine	10 - 25 mg/L 25 - 50 mg/L 50 - 100 mg/L 50 - 200 mg/L	Add 10g/L (1%) of gelatin to nearly boiling water and stir well to form a gel solution. Add very slowly to wine while it is still warm (before it solidify) with constant stirring. Rack off after one-two weeks. 1cc of the solution contain 10mg of gelatin.
Kieselsol	In combination with gelatin for white wine clarification and bitterness reduction.	5 times (in ml) of the gelatin conc. (in grams). e.g. 50 ml/HL for 10 gr/HL of gelatin.	The commercial agent comes in a 30% solution prepared for direct use. Addition one day prior to gelatin addition is recommended.

(continues next page)

(continues)

Agent	Use for	Dose range	Preparation and use
Isinglass	Clarifying white wine to brilliancy.	10 - 50 mg/L	To 10 liters of cold water add 10 gram of tartaric acid, and 4.4 gram of potassium metabisulfite. Mix until dissolved. Add 100 gram of isinglass while mixing, and let to stay for one or two days With occasional mixing. Rack off after one-two weeks. 1cc of the solution contain 10mg of isinglass.
Casein	# Bitterness reduction in white wine. # Over-oakiness in white wine. # Browning and pinking reduction.	50 - 250 mg/L of potassium caseinate; or 2 - 10 ml/L of low-fat milk.	Add 20 gr/L (2%) of potassium- caseinate to warm water and stir well. Leave over-night and repeat stirring. Good for one - two days.
Egg-White	Softening and polishing red wine.	1 - 2 egg-whites per barrel (225L).	Collect the egg-whites of fresh eggs and dissolve in salty water which contain 0.5%-0.9% table salt. Mix with a mixer and use freshly made.
PVPP	# Color reduction (browning and pinking) in white wines. # Stabilizing white and blush wines towards browning. # Reduction of bitterness in white and red wines.	100 - 700 mg/L in white wines. 100 - 200 mg/L in red wines.	Direct addition to wine with well stirring. Or mixing with wine (or water) to make 5%-10% slurry solution prior addition to the tank.
Carbon	# Deodorizing off-odors # Decolorizing browning and pinking in white wines.	50 - 250 mg/L	Adding directly to the wine. Settles in few days. PVPP helps to settle it down.

D. Filtration

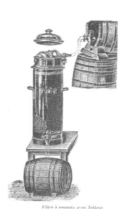

Filtre à amiante avec flotteur.

Cloudy wine will clarify if let to stand for a long time. However, even a wine which looks clear after a long time of settling, contains enormous amounts of particles in the microns (10^{-3} mm) range, including yeast cells, bacteria, pigment particles, proteins, fining particles, tartarate crystals, pulp and other components. The purpose of filtering is to facilitate the removal of most of these particles from the wine.

Filtration is done by two basic mechanisms; first by absorption of the particles on and in the surface of the filter texture, by electrical or cohesion forces. These filters are made of fiber pads or a composite of fibers and mineral particles. These filters are called therefore, *pad filters*. The wine particles are trapped in these filters on the whole volume of the pads surface area. The second mechanism is by size control of the pores, which prevent any particle bigger than the filter pores to get through. These filters are called *membrane filters* and are used mainly for very fine and sterile filtration.

1. Pad filters

The pad filters are made of fabric or paper pads, which are mounted tightly in a stainless steel or plastic frame (see figure). The mechanical set which is called a plate filter, contains a series of plates that hold the pads in between (one plate, one pad). The common dimensions of the commercial pads are squares of 20x20 or 40x40 centimeters. The pads are not identical on both sides. One side is rough, facing the inlet wine flow, and the other side is smooth, facing the outlet wine flow. The pads must be packed alternately between the plate filters (see figure).

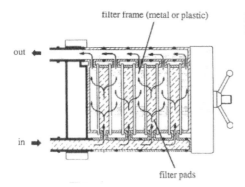

filter frame (metal or plastic)

out ←

in →

filter pads

Frame filter device. The wine flows through the frames, leaving the solid particles inside and on the pads.

So, when placing the pads in the plate frame, one should be aware of maintaining the proper order of the pads, to avoid having tiny particles and fibers of the pads enter the wine.

The fineness of the filter texture is responsible for two parameters: the distribution of the filtered particle sizes that can pass through the filter, and the resistance of the filter to flow. The particle sizes in wine which have to be filtered, range between sub-micron size (bacteria), up to tens microns (lees and fining particles, tartarate crystals). The commercial pad filters are numbered in a series according to the fineness of the filter. The higher the number, the finer is the filter. Sterile filters are marked by letters (EK, EKS etc).

When wine flows through a filter, the rate of filtration (volume filtered per time unit) is proportional directly to:

(1) The pressure difference between the input and output of the filter.
(2) The filter area.
(3) Inversely proportional to the viscosity of the liquid.
(4) To a factor called "resistance to filtration." The last term depends on two sub-factors:
    (a) The filter's resistance to flow of clear water, an inherent characteristic of the filter texture (pore size and packaging tightness).
    (b) On the filtered liquid viscosity.

The above parameters can be expressed in a formal relationship:

$$\frac{dV}{dt} = \frac{dP \times A}{\gamma \times R}$$

114

where: dV/dt – is the volume of filtered wine per unit time (filtering rate).

dP – is the pressure difference across the filter.

A – is the total filter area (pad's area x number of pads).

γ – wine viscosity.

R – filtering resistance.

The flow rate of typical commercial filters are given in the table:

Filter mark	20	30	40	50	60	70	EK	EKS
Liter/hr/m²	6600	6000	4600	3750	3400	2500	1100	750

Flow rate of water through filter pad at pressure difference of 10 p.s.i. The flow rates are in liters per hour, per one square meter area of the pads (the area of one regular size pad of 40x40 cm is equal to 0.16 m². The above flow numbers are therefore for about 8 papers of 40x40 cm.

After the plate filter is set and the pump has been connected, it is very advisable to flush the whole filter setting with 0.5% tartaric acid or citric acid, for a couple of minutes (in close cycle) in order to avoid introducing the "papery" flavor of the filter pads into the wine. The filter pads, the hoses and pump should be emptied from the solution after the flushing, before connecting to the wine tank.

The filtered particles are accumulated on and in the pads, causing its flow resistance to become higher, and as a result, pressure is built up across the filter surface between the input and output of the filter. The increasing resistance gradually reduces the flow rate on one hand, and as the pressure is built up, it can cause a break in the filter texture, leading the filter to lose its filtering capability. Therefore, this differential pressure should not be higher than about 2–3 bars (30–40 p.s.i.). When it reaches that pressure, it is recommended to replace the filter pads.

Another important factor, which must be taken into consideration, is the aeration of the wine during filtration. To minimize oxidation, especially in white wine, it is highly recommended to add sulfur dioxide to the wine during filtration. In the first filtration, after bentonite and cold stabilization, there is another option, that the SO_2 should be added before the bentonite addition and cold stabilization (for details see the previous section on stabilization).

When the second filtration is done (bottling filtering), the sulfur dioxide may be added to the wine a couple of days before filtering. After cold stabilization, a course filter can be used to clear the wine turbidity. Filter marks 30 to 50 allow a moderate flow rate and will clear the wine fairly well. If after a short time, the filter is clogged (and the pressure gets too high), the filter should be replaced by a smaller filter mark.

Some wineries use filter aids (diatomaceous-earth or kieselguhr), which are very small particles of minerals with a very large surface area, to prevent clogging the filter pads. The powder of diatomaceous-earth (DE) is mixed with the wine just before filtering and when flowing through the filter, it builds up layers on the pads containing the DE and the wine particles. The multi-layers of DE ease the flow rate. The amount of filter-aid, which is used for that purpose, is in the range of 50–100 gr/HL, depending on the concentration of the particles in the wine. In certain filter devices, the filter-aid can be administered into the wine by a dosing unit mounted at the input of the filter. But in order to use the filter-aids in filtration, special frames designed for that use should be available. The regular pad frames are not suitable for use with filter-aids.

2. Membrane Filters

These filters are made of synthetic polymers (cellulose esters) that have uniform tiny holes at the micron range. The membrane filter is a *surface filter*, in contrast to the pad filter, which is a *depth filter*. In the membrane filter, the particles are blocked on the surface by the size of the holes, which specify the filter characteristic. The common sizes in membrane filters are 1.2, 0.65 and 0.45 microns. The bigger size will block most of the yeast cells and the finest size will block most spoilage bacteria. The membrane filter is assembled as a plastic cartridge mounted within a vertical stainless steel tube. The wine flows into the cartridge, leaving the microscopic particles on its outer surface. During operation the accumulation of particles on the surface increases the resistance and causes pressure to build up. As mentioned above, the pressure gets higher, and the flow rate decreases. At high pressure, the filter may be permanently damaged losing its sub micron filtering capability. The maximum working pressure is about 3 bars. When reaching this pressure, filtration should be stopped, and the filter cleaned. In order to clean the membrane filter, water flow from the opposite direction (from input to output) will remove the particles, which are left on the membrane's surface. The membrane filter is generally used only before bottling the wine. At bottling, because the sterile filtering is the major measure taken for biological stability, and in sweet or off-dry wines it is sometimes the only assurance against re-fermentation, it is most important to check the filter before each bottling day (and after too!). The checking is done by the so-called "bubble point" test. To run the test, water is flowing first through the filter to fill it with water. Then a nitrogen line from a cylinder is connected to the inlet of the filter and the outlet is connected by a thin plastic tube to an open vessel containing water. The main nitrogen valve is then opened and by using the regulating valve the pressure is slowly increased until bubbles of nitrogen appear at the filter outlet (in the water). This pressure is needed to force the bubbles of gas to act against the water surface-tension across the filter's holes. The smaller the radius of the hole, the greater is the pressure needed. If there is a break in the filter texture, the bubble pressure would be lower than the normal test pressure. For 0.45m cartridge (the common filter for white wine bottling) the bubble point pressure should be between 25–40 p.s.i. If the bubble point appears at lower pressure, the cartridge should be replaced.

3. Mechanical Separators

There are two devices which can be used to clear turbid wine:

One is a continuous diatomaceous-earth filter (Kieselguhr) which is based on building a layer of small irregular DE particles like a filter texture on a permanent support, preserving these filtering layers with a constant flow of pressure. The advantages of the diatomaceious-earth (DE) continuous machine are: (1) high flow rate, (2) very high filtering capacity before the filter gets stuck, especially when the solid particle concentration is high (in must or young wines), and (3) economical advantage compared to pad sheet filters (DE is much less expensive than filter pads). The disadvantages are: (1) Complicated operation, (2) Long preparation time before starting the filtering (about 1- 2 hours), and (3) Operational error may lead to the necessity to start the whole preparation from the beginning. The amount of DE, which is needed for building the "cake" before filtering, is about $1.5 – 2.0$ Kg/m^2. During filtering, the DE is continuously fed into the wine by a dozer from the

DE container in the machine. There are different particle sizes composing the DE, suitable for the different filtrate particles. For must or young wines, Celite 535, 545 can be used. For clean wines, Celite 501, 503 are preferred. During preparations, before filtering the wine, the "cake" should be washed (in closed cycle) with 0.5% tartaric acid or citric acid (to remove the "papery" taste), and then with clear water. The initial differential pressure before starting the filtering is about 1 bar. The working pressure can rise up to 6–7 bars and then the filtering should be stopped, the machine has to be cleaned and be prepared from the beginning. The operation of this filter device is quite complicated and sensitive, and to our opinion it is suitable for large quantity of wines, namely, to medium size wineries and up.

The other machine for clarifying must or wine is the centrifuge, which can clear turbid wine easily and quite satisfactorily, but its price is high and perhaps not economical for a small scale winery.

• • •

And final remark on filtration pumping: One should use only continuous pressure pump (e.g. centrifugal pump or oval one) for the filtering operation, rather than pulsed pressure pump (e.g. a piston pump or air membrane pump), which generates on the filter a cyclic pressure rather than a continues one.

E. Blending

Blending of wine is done in order to achieve different goals:

(1) Overcoming certain deficiencies or defects;
(2) Balancing the wine; and
(3) Enhancing complexity.

Blending can be done at any stage of winemaking, from must blending before fermentation, or at any wine-processing stage afterwards. The corrections made by blending two wines together, enhances both of their respective characteristics; one may be over endowed with certain character and the other may suffers from a deficiency of it. By blending them, both wines will improve in that specific character. However, there are many other parameters involved in wine quality and all of them have to be considered. The "art" of blending is not a simple one.

Blending can be done with different varieties, different vintages of the same variety, different vineyards or locations of the same variety, and different lots of the same vintage (tanks or barrels).

Each country has its special rules concerning the blending and labeling of wines. In California for example, in order to state on the label a varietal name, it is mandatory to blend at least 75% of that specific variety in the wine. If the blending is such that there is less than 75% of any variety, the wine has to be labeled as generic wine. This does not, however, mean that a generic wine cannot be an excellent wine.

As for the region of origin (appellation), the approved American Viticulture Appellation requires that 85% of the grapes must come from the indicated region in order to be stated on the label. The regions do not necessarily coincide with geographic boundaries.

For vintage statement, 95% of the wine must be from the year that is printed on the label.

The alcohol content in table wine should be in the range of 7%–14%. The winemaker should be aware of all regional rules when blending his wines. The same principles of specific regulations are practiced in other countries as well.

In varietal blending there are certain conventions regarding which varieties are suitable to blend. For example, Cabernet Franc and Merlot are commonly blended with Cabernet Sauvignon, or Semillon with Sauvignon Blanc. French Colombard is a good blend with any fruity white wine for its high acidity content and floral aroma. In certain cases of generic wines, there may be up to four varietal blends with excellent balance and quality.

In regular cellar operations, it happens quite frequently that certain lots of the same vintage develop differently—better or worse—than the average of that vintage. In such cases, the question is whether to separate these lots (if better, to separate it as special reserve; or if worse, to sell it as bulk), or to blend them together to get the best from what's at hand (professionally and economically). No formula or recommendation can be made. The decisions are ad-hoc. Also if a press run is left separated, then at blending special attention has to be paid to it, because it may contain high concentration of tannins, high pH, volatile acidity and a flat taste.

The parameters that can be easily corrected by blending are: acidity, pH, alcohol, color, tannin, varietal aroma, freshness and fruitiness, oak flavor, volatile acidity, residual sugar, bitterness and off-flavor. Some of these parameters are in excess or in deficiency in the wine. The blending may "correct" that parameter to the desired level, or at least make it tolerable.

Some of these parameters are quantitatively measurable (like acid, residual sugar, alcohol, volatile acid, color) and their concentrations are linearly related. Other parameters, such as aroma, flavor, off-flavor, can be appreciated mainly by organoleptic judgment.

For measurable parameters, the volume ratios of the wines to be blended, according to the desired concentrations of the parameter in question, is given by the simple formula:

$$P_1 + XP_2 = P_b(1 + X)$$

where: P_1 – concentration of the actual parameter in wine No.1.

P_2 – concentration of the actual parameter in wine No.2.

P_b – concentration of the actual parameter in the blended wine.

X – volume ratio of wine No.2 to the volume of wine No.1; ($X = V_2/V_1$)

For example, wine No.1 with residual sugar of 0.8% is blended with wine No.2 which has 3.8% sugar, to obtain a designed blend with 1.8% residual sugar. Then $0.8 + 3.8 X = 1.8 (1 + X)$, with a solution of $X = 0.5$. Thus, by taking 1 volume of wine No.1 and 0.5 volume of wine No.2, the blend will contain 1.8% residual sugar.

For those parameters, which are not directly measurable, a process of trial and error must be carried on to find the best result. All blending tests must be tried first in the laboratory. The aid of close colleagues is highly recommended. When the attempted formula has been concluded, a small sample of five gallons has to be blended and left alone to, what it is called, "get married." This *"marriage"* takes a couple of weeks, is then tasted, and if satisfying, the whole batch of wine can be blended accordingly. If there is some doubt about the exact volume ratio, the five-gallon test can be repeated with some small variations, and after the "marriage" period, a decision can be made as to which is the best one.

After blending the wines at an advanced stage of processing (e.g., before bottling), sometimes a stability problem might arise, especially regarding precipitation. For this reason, it is best to blend very early, before stabilization takes place. On the other hand, only when a wine is in a stage of maturity can one evaluate the full character and quality of the wine. It is also our opinion that after blending is completed, the storage temperature of the blend should be 5°C–10°C. This temperature should be maintained for a couple of weeks to check for instability. If there is any, for whatever reason, care should be taken before bottling.

F. Maintenance
1. Sanitation
Besides its aesthetic value, winery sanitation is most important in preventing and minimizing bacterial and fungal spoilage of wine. Although wine is less susceptible to harmful spoilage than other food products, it is still sensitive to a number of microorganism infections, which may affect its quality considerably. Winery sanitation includes cleaning and sanitizing all the equipment in the winery that comes in contact with the wine, from crushing to bottling. Of all the many materials and cleaners on the market, we choose to emphasize here those that are most efficient and most used in the wine industry.

Water: Can be used cold, hot and as steam. Cold water cleans by dissolving and physical removing the soil from the surface. Hot water is much more effective than cold water. At 80°C the water has a double effect, that is, physical cleaning and sterilizing at the same time. Three major devices are in use for applying water cleaning in the winery:

(1) Water guns use water system pressure. Spraying water with a water gun or with an open hose is the most common and basic technique for equipment cleaning in the winery. During harvest time, at the end of any working day, all the machines that have been used (hopper, press, drainer, destemmer/crusher, hoses, pumps, tanks) should be washed with water to remove all residues left over.

(2) The sprinkler ball, which can be operated by pump pressure in a close cycle or by the water system pressure, is used inside containers, like tanks and barrels, to spray the water uniformly on the container walls.

(3) Hot water machines are one of the most useful tools in the winery. It is connected to the water line and can heat the water instantly up to about 80°C and eject it at very high pressure through a gun. The cleaning ability of this machine for any surface (metal, cement, wood, plastic) is most effective. It also, very effective, conserves water.

Besides cleaning, hot water can be used as a sanitizing agent against almost any microorganism. Its effectiveness depends on the temperature and the time of application. At 80°C, 10 minutes of hot water contact would kill almost any bacteria and fungi (but not all spores). For example, the filling machine in the bottling line has to be sanitized at the beginning (and end) of every work day by passing through hot water at 80°C for 10–15 minutes.

Alkaline solutions: The strongest agent in this category is sodium or potassium hydroxide. Its main cleaning function is to saponify oily soil on the surface and to emulsify it so it can be removed by water stream. Milder but still alkaline in its activities is sodium carbonate (soda ash). Alkaline solutions are also used to dissolve the hard layer of deposited tartarate on the inside walls of tanks and barrels or any equipment that has come in contact with cooled wine or must for an extended period of time. Alkaline solutions are used at concentration of about 0.2%–0.5% in water. When the solid hydroxide is dissolved in water, it releases heat, and care should be taken whenever this agents is used. Alkaline solutions are skin irritating and very dangerous to the eyes. Glasses and rubber gloves are the minimum safety measures when working with alkaline.

When using alkaline, the general practice is to prepare the solution in a small container of 100–200 liters, and apply the solution to the equipment (such as tanks) by spraying it with a sprinkler ball in a closed cycle. The cycle contains the open solution container, the pump and the tank, which is connected to the container by a hose. When cleaning, the tank's opening should be closed for safety reasons. About 10–20 minutes of application is usually enough to remove any sediment from the tank's walls. To clean small equipment, immerse it in the container for a couple of minutes. Barrels are best cleaned (if necessary) with a less potent alkaline soda-ash. They should be filled with the solution for 15–30 minutes, and then emptied and washed.

The residue of alkaline on the surface is difficult to remove by water. Therefore after any application of alkaline the surface should be rinsed with water and washed with 0.1% citric acid solution (which interacts with the alkaline surface residue), and then finally rinsed with clean water to wash away the acid.

Chlorine solutions: Based on Sodium or Calcium Hypochlorite [$NaClO$, $Ca(ClO)_2$] which in solution can be in two forms: hypochlorite anion (^-ClO) and hypochlorous acid ($HClO$), depending on the pH. At low pH when the acid is predominate, the solution acts as an excellent sanitizing agent against most microorganisms. Its action is based on its ability to penetrate the microorganism's cell wall and chlorinating the cell's enzymes. At high pH, when the hypochlorite anion is the main form, it has a powerful oxidation potential, which can bleach organic stains, and cause it to emulsify and be removed by excess water.

The commercial reagents are sold as bleaching agents (hypochlorite + alkaline), or as chlorinating agent (hypochlorite + acid). The former is usually sold as powder and the latter as liquid. The powder form has an effective concentration of about 5 gr/L and the liquid quantity depends on its initial concentration. In any case the exact producer's instructions should be followed for good cleaning effect and for safety awareness. Here again, as with the alkaline solutions, if the bleaching powder is used (alkaline), it has to be followed by washing with 0.1% citric acid and water.

Iodine: An excellent sanitizer. It consists of a solution of iodine in water (plus potassium iodine to increase solubility), and surfacetant materials to increase its surface contact. The pH is about 4 by phosphoric acid. This solution is active against microorganisms due to the free iodine. The brown color of the solution makes it easier to follow its removal with rinsing water after use. The effective concentration is about 25 mg/L, but the exact quantity should be read on the producer's instructions.

2. Storage

Wine can be stored in different kinds of containers: wood, concrete, iron, plastic and stainless steel. When concrete and iron are used, the interior of the containers is covered usually with glaze, ceramics or polymeric paint. In modern wineries, wood, concrete and iron containers are not used anymore; stainless steel has replaced them for ease of maintenance and its inert surface. Plastic containers, which are cheaper than stainless steel, are rarely used in the wine industry, for no specific reason. Probably in the long run their amortization is higher than stainless steel containers. They're also harder to clean. Oak containers (barrels and casks) will be discussed in a separate chapter.

The cleanliness of containers (or tanks) is a basic requirement in cellar operation. This means that immediately after racking, the tank should be washed, brushed if necessary, and left empty and clean for its next filling. Before filling any tank with wine, it is highly recommended to recheck the tank through its opening.

When storing white wine between different stages of cellar operations, the main objective should be to minimize air contact. The reasons are twofold: preventing oxidation and acetobacteria infection. In red wine, only the last reason is of major importance. To minimize oxygen in white wine storage, it is necessary to keep the tank full as much as possible, and also to fill the space above the wine with an inert gas, such as carbon dioxide or nitrogen. The gas lays on the wine surface as an inert blanket, protecting it from coming into contact with oxygen. Carbon dioxide is heavier than air and will stay on top of the wine, but it is highly soluble in the liquid. Nitrogen is much less soluble, but it is lighter than air and diffuses out. So, when using either gas, the top of the tank should be checked and refilled every month if necessary. The cover or top of the tank should be closed and air-locked to minimize gas diffusion and to prevent pressure differences between the tank and the outside.

As for the temperatures during storage: for white wine the preferred range is 8°C to 12°C and for red wines 20°C to 25°C.

Barrel Aging

Chapter V: Barrel Aging

Wooden cooperage adds an *extra cost* to wine production due to the high investment price, increased labor, much larger storage space, as well as losses caused by wine evaporation from barrels. In spite of all of these extra expenses, certain wines (red and white) must spend some period of time in small oak cooperage in order to finish their maturation before being bottled.

As a general rule, one of the most important parameters in red wine quality is the balance between the varietal aroma, the aging bouquet, and the oak character. The last two angles of this quality triangle are achieved by aging the wine in oak cooperage. Wooden containers are not easy to maintain, because of leakage problems, sanitation difficulties, lack of temperature control and problems of maintenance when not in use. They are also expensive. Aging the wine in oak barrels has two goals:

(1) *Slow Oxidation* of the wine, which enhances its bouquet, and softens its tannin by slow polymerization of its phenolic compounds. This function is fulfilled by the periodic topping operation, which is needed because of the evaporation of the wine through the wooden surface.

(2) Adding *Oak Phenolics* into the wine, by extraction from the inner surface of the barrel. These compounds are most compatible with the wine aroma component, and together they expand the complexity of the wine's bouquet.

A. Cooperage
1. Barrel Making
The basic information regarding the origin, making, characteristics and qualities of barrels will be presented in this section.

Origin: The factors to consider when choosing the tree species for wine storage are:

(1) Its strength.

(2) Large trunk diameter for economical production.

(3) Straight and clean wood with no defects or knots to enable clean-cut staves with a low chance of leaking.

(4) Good flexibility.

(5) No undesirable extraction components which might be transmitted from the wood into the wine.

Redwood fulfills most of these demands, but it contains too high a level of extractable phenols, which may damage wine stored over long periods. Red oak is too porous for reasonable wine storage. Other trees were found to have an objectionable flavor or color, which affects wine stored in their barrels. No less important is to have a *desirable* flavor extraction potential which is compatible with wine aroma. The ultimate wood, which meets all of the above characteristics is *white oak*. It has all the mechanical factors, as well as the desired aroma, which is the most important factor.

A cross section of a tree is shown in the following figure:

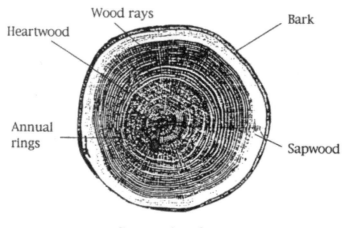

Cross section of a tree

Its features are (from outside to the center): *Bark,* which has two parts, the outer corky dead one, and the inner living part. Next to the bark is a light colored section called the *sapwood,* which is the wood tissue that carries sap from the roots to the leaves. The sapwood is softer and very permeable, and therefore cannot be used for barrel making. The next inner section, which is darker and harder, is called the *heartwood* and its function is to give the tree its mechanical strength. Only the heartwood is suitable (after drying and seasoning) as cooperage material.

The oak genus *Quercus*, has about 300 known species. There are basically two groups of oak trees, *red* and *white* oaks, and only some of the white oaks are suitable for cooperage. The main European oaks are *Quercus robur* and *Quercus sessilis*. The two are considered to be different varieties of the same species. European oaks are distributed all over Europe, mainly in France, Italy, Portugal, Yugoslavia, Russia, and Scandinavia. *Quercus robur* is also called *Q. pedunculata*, while *Quercus sessilis* is sometimes called *Q. sessiliflora* or *Q. petraea*.

The main American oak is the *Quercus alba* species, grown mainly in Kentucky, Missouri, Kansas, Oklahoma, Arkansas and Texas. Besides the species difference, different geographical regions produce variable kinds of woods, which are different in their physical and chemical characteristics. European oak is definitely different in its flavor from the American oak, and between the two European varieties, *Quercus sessilis* is more tannic and contains more extractable solids than *Quercus robur*. These two varieties are not always separated, and in many forests they grow side by side. Because France is one of the major producers of wine (and brandy) barrels, we will pay some more attention to French oak. The main regions where oak is grown are *Limousin, Trancais, Allier, Nevers*, and *Vosges*. It is very difficult to assure the origin of the oak because oak logs pass so many hands, and are mixed in the wood mill by those who cut them into staves. Instead, it is common to separate the staves according to the coarseness of their grain (fine, medium and coarse). The very coarse are dealt as Limousin, the medium as Nevers, while the fine grain as Trancais, Allier and Vosges. Analytical assignment of the composition profile of 11 major components in oak wood was recently studied, with a meanwhile conclusion that American and French oak can definitely be distinguished from each other, where it is not so clear between the French regions themselves. More studies are needed to verify and establish the differences between French and American oaks.

Barrel Production and Characteristics: Before describing its production, let us be familiar with some barrel terms shown in the following figure:

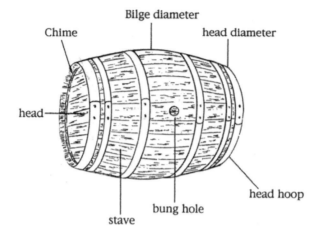

First, the trunk is sawn to logs of the desired staves length. In the traditional French barrel making, the staves are cut by radial *splitting* (with an hydraulic wedge or an axe) the cross section of the logs, to "quarts" (split sections) along the natural rays of the wood. This is preferred to the other option of *sawing* the cross section, which does not follow the natural wood grains. Sawing is very common in the American barrel industry, and it utilizes much more of the wood material (less waste) than the splitting method (by almost twice). Splitting is more expensive than sawing because it is a hand operation, and it also produces more waste material than exact cutting with a saw. The sapwood is cut off from the split quarts and only the heartwood is used for making the staves. The quarts (split) are then cut (now by sawing) to staves, so that from each split one to four staves can be produced.

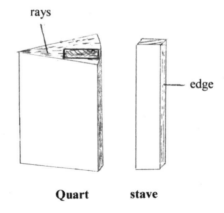

The wood rays are then parallel to the stave's surface (perpendicular to the barrel radius), which otherwise would be along the radial direction of the barrel. Because the wood rays run along the staves, in many cases when the barrel is leaking, the leak is found on the chime of the barrel and not along its side (the diffusion takes place all the way out through the stave length). With the structure of the staves in the barrel, the leaking is very much less than if it would leak straight along the stave width (which is only 22–28 mm thick). Also the rays being parallel to the staves contributes to their flexibility and bend ability when heated to form the barrel.

After cutting, the staves have to be dried, either in the open air, or by kiln drying (mostly for whiskey barrels). Open-air drying (exposure of the staves to rain and sun) may take from eighteen months up to three years. The desired moisture content of the wood, when seasoning is completed, is about 16%–18 % water. If the wood is not dry when the barrels are made, they will shrink later and will not hold wine without leakage. The seasoned staves are then shaved to the exact tapered shape so as to fit in place when assembled into a barrel. The staves are assembled together (held with six to eight hoops), bent by the aid of open fire and mechanical tools, heads are placed at both sides, a bung is drilled in the bilge of the barrel, and a final polishing of the surface is given to the new born barrel.

Cooperage containers can be made in various shapes. The most common and easiest to build is a cylinder whose *bilge* (the maximum diameter at the bung location) is about 1.2 times the *head* diameter (the minimum diameter of the cylindrical shape), and the length of the barrel is about 1.3 its bilge diameter. The extraction rate of the barrel's phenolics into the wine depends on the inner *surface to volume* ratio of the barrel. The following figure shows this ratio as a function of the barrels volume in a cylindrical barrels with the above relations of diameter and length:

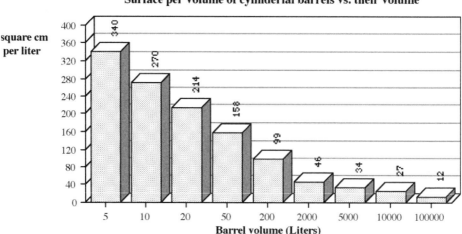

Surface per volume of cylinderial barrels vs. their volume

The surface/volume ratio at two different volumes is related very closely according to the formula

$$[S/V]_2 / [S/V]_1 = (2.15)^{\log V_2 / V_1}$$

Where: S/V is the surface to volume ratio, and V_1, V_2 are the volumes of two different barrels.

For example for $V_1 = 10L$ and $V_2 = 10,000L$, the S/V ratio is:

$$(2.15)^{\log 10000/10} = (2.15)^{\log 1000} = (2.15)^3 = 10$$

as can be verified from the above figure (270 cm^2/L and 27 cm^2/L respectively). The formula is general for any volume ratio. The decision about which volume is most suitable for wine aging has to take into account the storage time needed at each volume, the handling and maintenance effort for each barrel size, and the basic cost per wooden volume. Practically, the most common barrel size in winemaking is the 225L (60 gallon) French style, called *barrique*, and the 190L (50 gallon) American style. Other common sizes are 300L, 350L, 400L, and 500L.

Barrels are built in two major shapes, namely *Bordeaux* and *Burgundy* styles. The *Burgundy* style is somewhat wider and shorter than the Bordeaux one. Each style has two types: the regular type which is called *Bordeaux-chateaux* and *Burgundy-tradition*, and the massive (for transportation) type, which is called *Bordeaux-export* and *Burgundy-export* .

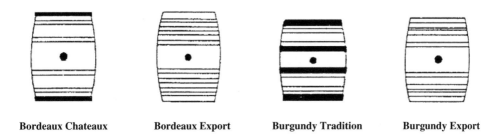

| Bordeaux Chateaux | Bordeaux Export | Burgundy Tradition | Burgundy Export |

Also in many cases the *chateaux* and *tradition* types have pinewood cut on their head hoops. The maximum diameter of 225L barrels (at the bilge) is about 65–70 cm, and the head diameter is about 54–57 cm. The length of the barrels is 88–92 cm. The staves thickness in the *Chateaux* and *tradition* types barrels is 21–24 mm, where that of the *Export* types is about 24–28 mm, and therefore they are heavier than the thinner types by about 5%–15%. A typical dry barrel weighs about 45–55 kg. When the barrel is wet and saturated with water (or wine) its weight increases considerably.

Barrel Toasting: It is very common to toast the inside surface of the barrel before the heads are placed in position. The toasting is done with an open fire of small oak cuts at three levels of toasting: *light, medium* and *heavy*. It is most important that the toasting be done gently, at even intensity over the whole surface, so as not to burn it to charcoal. *Light* toasting is done just on the surface with no penetration to the inside of the wood. *Medium* toasting is more intense and penetrates into the wood to about 2 mm from the surface, while *heavy* toasting goes as far as 3–4 mm deeper.

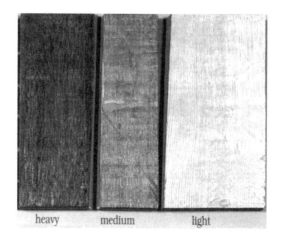

heavy   medium   light

In whisky (or bourbon) barrels it is common to ignite the inside surface and let it burn for few minutes, which causes a film of black charcoal to develop on the inner surface. Such toasting is called "charring". Back to wine barrels, in many cases the heads are also toasted before being placed in position. The primary reason for toasting is to soften the phenol extraction of the new barrel by decomposing some of the gallic and ellagic tannins, as well as the polysaccharides near the barrel surface. Also, the heat decomposes the wood lignin to release some volatile phenolics (see later in this chapter, and in the General Aspect chapter section D). The toasted barrel gives a very distinct aroma described as 'fresh bread', 'butterscotch', and 'toasted almond', to the wine being stored in a newly toasted barrel, mainly to white wine. Another benefit of toasting, from the mechanical point of view, is that it relieves some of the stress of the newly made barrel, thus reducing the cracking potential of the weak bent staves. Very useful information on cooperage in general, and the wine-cooperage relationship can be found in the literature.[1, 2, 3, 4, 5]

2. Barrel Evaporation

It is well known that wine stored in barrels does evaporate, and continuous topping is necessary to keep the barrel full. Water and ethanol diffuse into the wood, carrying with them some nonvolatile components. A cross section of old staves from a red wine barrel, shows red pigment in the inner side of the staves up to a few mm. The penetration of nonvolatile compounds is limited in depth, but the volatile water and alcohol can diffuse all the way out from the barrel to be lost in the cellar environment. The evaporation rate depends on the temperature (higher vapor pressure at higher temperature), air movements, and on the relative humidity in the cellar. Water molecules are smaller than ethanol molecules, and are therefore more able to diffuse out. But on the other hand, water vapor pressure is lower than that of ethanol. In general, under typical barrel storage conditions, the total evaporation rate is about 2%–4% per year, which is about 4L–9L per 225L barrel, annually. If the bung is tightly inserted, the diffusion of water and alcohol can develop a vacuum inside the barrel. The reduced pressure may develop to about a few inches of water. The existence of vacuum in the barrel is a very good proof that air is not able to diffuse into a wet full barrel, which means that oxidation of the wine through the barrel walls is not significant. The oxidation of wine stored in barrels happens during cellar operations such as barrel racking and topping. Also, being a small volume container, any topping, which is done frequently, is enough to add the desired oxygen for the barrel-stored wine to age. The evaporation rate of red wine (Bordeaux varieties) stored in the same barrels (French barrels, Allier oak, Chateaux style, medium toast) in two storage conditions: (1) Natural conditions of a cave, and (2) An air-conditioned cellar, was studied. Temperature and relative humidity measurements showed that they were very stable in the cave (during the 50 weeks of the experimental period), while in the air-conditioned cellar, the fluctuations of both parameters were drastically higher. The average temperature and relative humidity in the cave and in the air-conditioned cellar were 16.7°C / 92.5%, and 13.5°C / 74% respectively. The mean volume of topping wine (after correction for expansion/contraction changes at each topping) was 2.5L/barrel per year in the cave, compared to 7.0L/barrel per year in the air-conditioned cellar. It looks that the stable conditions existing in a cave help reduce wine evaporation. If a stable controlled air-conditioned cellar could be maintained, probably the evaporation rate would be the same as in a cave.

In regard to the difference in evaporation rate between water and alcohol, the final concentration of ethanol (in the above experiment) in the wine kept in the cave dropped by 0.2%, while the air-conditioned cellar wine gained 0.1% (from 13% before the experiment). It looks therefore, that the balance point, namely, equal evaporation of water and ethanol from wine kept in barrels must be somewhere above the 74% relative humidity (the mean value in the air-conditioned cellar). This finding is somewhat different from a previous report that below 60%–65% relative humidity, water evaporates faster then alcohol, and above that humidity, alcohol evaporates faster. In the first case (higher water evaporation) the wine became more concentrated with alcohol, and vice versa in the second case. Another study confirmed the higher losses of wine under ambient conditions compared to controlled environment (temperature and humidity). The experiment was done with *topped* barrels (weekly topping) and *rolled barrels* (bung at 2 o'clock position with no topping for the entire length of the experiment). From the experiment it is evident that losses are much higher in the uncontrolled conditions (temperature variations between 14°C–26°C and humidity between 54%–90%), then in the controlled ones (14°C–16°C and humidity between 85%–95%). The alcohol changes were like those of the previous study.

3. Barrel Maintenance

Preparation of New Barrels: New barrels should be swelled properly in order to avoid any damage to the wood. Some guidelines might be helpful:

(1) Place 20 to 30 gallons of cold water in the barrel (horizontal or vertical position), wet the whole inside and let swell overnight. The barrel might leak and lose part of the water if the staves or the heads were loosened after a long period of being dry. In such a case, fill the rest of the barrel with water, and if the leak doesn't stop after a day, continue to add water for a few days until the leaking stops.

It may help to stop the leakage by inserting a thin piece of oak into the leaking place with a hammer. If the leakage has not stopped, empty the barrel and put new water in for another few days. Do not use hot water for swelling because the inside of the barrel will swell faster then the outside, and the wood may crack (especially the bung stave). If a new barrel still leaks after a week of soaking, it needs a professional cooper to repair it.

(2) Do not use any chemical substance except clean water in preparing a new barrel. It may change the natural flavor of the newly toasted barrel. After the barrel is swollen, you may use warm water (not hotter then 40°C).

(3) After the barrel is watertight, before putting wine into it, empty the water and let it drain (bung down) for an hour.

Storage of Used Empty Barrels: In general it is always better to keep the barrels full with wine. When this is not possible, care should be taken to store empty used barrels:

(1) Rinse the barrel well with water after wine has been removed. Let the barrel drain, add about one liter of water, and spray sulfur-dioxide gas inside the barrel for 3–4 seconds, and close the bung with a silicone stopper. Burning sulfur paper inside the barrel is another sulfiting alternative. Cycles of rinsing, draining, adding water, and sulfiting every two-three months are necessary to keep the barrel in a good shape (mechanically and microbially) for later usage.

(2) Before wine is placed in a used empty barrel, it has to be swelled as a new one. The water placed in the barrel during the swelling also removes the accumulated sulfur-dioxide from the barrel interior.

(3) In the rare case when a barrel has not been used for a very long time, and the hoops have loosened, do not force the loosened hoops into a new position (further up the barrel curvature) to tighten the staves. If you do so, when the barrel swells there is a good chance some staves will crack. Instead, wet the barrel patiently by adding water frequently, until it swells enough to hold water. Only then should you fasten the hoops more or less to their original positions.

Using Chemicals in Barrel Maintenance: This is not recommended, because chemicals may leach out the barrel flavor, and may also change the grain structure of the wood. They should be used only in severe cases of microbial spoilage, which caused undesirable odors and flavors in the barrel:

(1) The most drastic chemical for such treatment is caustic soda (NaOH) or soda ash (Na_2CO_3). The concentration should be between 100–400 gr/barrel of 60 gallon (0.5–2.0 gr/L). Do not use higher concentrations. Dissolve the caustic soda in water, and add to the barrel when it is half full of water. Then add more water to fill the barrel. After a few hours, empty the barrel and rinse with water several times. Neutralize the residual caustic soda with citric or tartaric acid (100 gr/barrel). Caustic soda is highly basic (pH above 13), and care should be taken when dealing with this chemical. Especially avoid any contact with the eyes.

(2) Chlorine treatment (or iodine) is also very effective in sterilizing microbial organisms. The recommended concentration is 200 ppm of active chlorine (or 15–25 ppm of iodine). There are several chemical sources for chlorine (and iodine) and the user should look for the producer instructions. The chemical has to be dissolved in water before adding it to the barrel. It should stay in the barrel for a day or two. Then the barrel should be rinsed with water several times, and be neutralized with citric or tartaric acid. There are some questions about using chlorine to sterilize barrels, because of the possible reactions between it and the barrel phenolics, which may lead to undesirable, off-flavored compounds.

(3) Citric or tartaric acid is used to neutralize any basic residual chemical left in the treated barrel. The recommended concentration is 0.5 gr/L. It should be dissolved in water before being added to the barrel, then be left for a few hours. After removal, the barrel should be rinsed with clean water, then filled with wine or be sulfited for storage.

Tartarate Removal: When tartarate has been deposited on the inside of a barrel:

(1) It is best to remove it by soaking with warm water (40°C). Repeat the soaking several times until the warm water has dissolved all the tartarate. Using a high pressure spinning ball inside the barrel is also efficient for cleaning the hard layer of tartarate.

(2) If very heavy tartarate is deposited inside the barrel, caustic soda can dissolve it (see above).

Maintaining Full Barrels at Storage: Wine stored in barrels should be well maintained in order to minimize any possible damage to it:

(1) Due to wine evaporation, top the wine at time intervals that do not allow the added wine (at each topping) to be more then 0.5L–1.0L in a 60 gallon barrel. After filling wine into a new barrel, the first topping time may be quite short (a week or two). It then stabilizes for a longer period.

(2) If wine leaks, but not so seriously that it can't be kept in the barrel (which usually will seal itself after a while), clean the outside surface with water, so mold won't grow on it. A leaking barrel can also be repaired by pushing oak sticks into the hole. If the leak is serious and does not stop, it will be necessary to replace the barrel. (According to Murphy's Law the faulty barrel will be the lowest barrel in the stack, with several barrels above it).

(3) Wine stored in new barrels tends to extract oak flavor very fast. Frequent checking and tasting of the wine is strongly recommended (especially with white wine) in order not to over-oak the wine.

(4) Red wine stored in barrels should be held between six months and two years, depending on the wine quality, barrel age, and storage conditions. When white wine is aged in barrels, the length of time should be much shorter, from a few weeks up to several months. Storage time depends very much on the storage temperature. At higher temperatures the extraction rate is higher, so the storage time should be shorter.

Life Cycle of Barrels: The length of time, before the barrel exhausts its flavor is about four to six years. The length of time needed to age wine in a barrel, to extract the desired flavor, gets longer as the barrel becomes older. In many French Chateaux, each vintage is held in new barrels for a period of 18–24 months. Afterward, the barrels are sold.

(1) A good practice for starting the life cycle of a new barrel is to ferment white wine in it. The yeast softens the barrel tannin for the next refill. Also, the new toasted barrel gives the white wine a roasted almond flavor, which is very compatible, for example, with Chardonnay or Sauvignon Blanc wines. In some cases the wine is left over with the yeast (*sur-lees*) for an extended period of few months, and then blended with stainless-steel-fermented wine. In certain wineries in Bordeaux, Sauvignon Blanc is left over sur-lees in new barrels for one to two years, with occasional stirring of the lees. The yeast lees protects the wine from becoming over-oaked. Instead, it becomes very round and full-bodied, with a surprising fresh-fruity aroma and extremely complicated bouquet.

(2) Controlled temperature and humidity is most recommended for barrel storage.

More information on barrel maintenance can be found in many barrel producers pamphlets, and some literature papers.[6]

B. Barrel Aging

The use of oak barrels to age red and some white wine is well established. The overall effects of red wine storage in barrels are:

(1) Slow and controlled oxidation, which softens the wine's tannin and increases the red color intensity and stability, by condensation between anthocyanins and other phenolic compounds.

(2) Extraction of oak phenols and flavor components from the barrel, which enhances and expands the wine's complexity.

(3) Evaporation of water and alcohol at an annual rate of 2%–4%, which in fact increases the wine's dry extract concentration, and its flavor.

(4) The gradual development of an aged wine bouquet.

Some basic information on the composition of oak extraction is important for understanding the wine/barrel relationship.

1. Oak Components & Extraction
Phenolic extraction

Tannins: The most extractable phenolics from an oak barrel are the nonflavonoids phenols such as *gallic acid*, *ellagic acid*, and *lignin*, along with their degradation products. The total phenols extraction was measured in new French and American barrels filled with white wine (Sauvignon Blanc) [12f].

The wine was aged in the barrels for about three months. Then the barrels were refilled with new wine for the same period. The quantities of phenolic extractions in the two series of barrel aging are shown in the following figures:

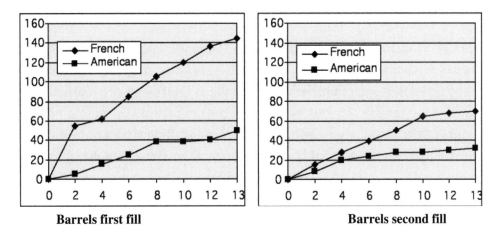

Barrels first fill   **Barrels second fill**

Phenolic extraction from French and American barrels during first and second filling with white wine (in mg/L). Time scale in weeks.

It is very clear that the French barrels contain more extractable phenols than the American barrels. Notice also the difference in the extract concentration between the new and used barrels. In the second refill the French barrel extract dropped considerably, while in the American one, it is more or less the same. This pattern has been shown in other works as well [12a].

The phenolic compounds extracted from barrels can be classified into *nonvolatile phenolics,* which might contribute *astringency* to the wine, and *volatile phenolics,* which enriche the wine with distinctive *oak bouquet.*

Nonvolatile Phenolics: The astringency *thresholds* of oak phenolics in wine were recently studied. It was found that the astringency *thresholds* of seasoned oak extract containing *only* nonvolatile phenolics (e.g. gallic and ellagic acids), which was added to white wine and to model solutions, is: 800–1600 mg/L (as Galic Acid Equivalent) when extracted from French oak, and 200–400 mg/L from American oak. These findings were compared with the actual concentrations of phenolics extracted by white wines and model solutions aged in wood for 12 months. The *total* phenolic that was extracted during that period was in the range of 200–300 mg/L from French oak and

100–200 mg/L from American oak. The conclusion is obvious, that there is practically no significant addition of astringent phenolics to wine from barrel aging. Or in other words, barrel aging *does not add an extra astringency* to wine. These findings were also supported before.

Volatile Oak Phenolics: Known also as *Aromatic aldehydes*, they are the main degradation products of lignin. The degradation may take place either in the acidic wine medium, or thermally when the barrels are toasted. The major compound is vanillaldehyde (vanillin), which gives wine a typical vanilla flavor. Regular concentration levels of vanillin in oak aged wines are much above its taste threshold (which is 0.5 mg/L in 10% ethanolic solution). Syringaldehyde, coniferyaldehyde and sinapaldehyde are some other ex-tracted aromatic aldehydes:

vanillin	syringaldehyde	coniferyaldehyde	sinapaldehyde

The toasting level will determine how much such aromatic aldehydes will be potentially extractable from a new made barrel.

Aromatic aldehydes are formed also by lignin degradation in the ethanolic-acidic solution of wine.

A very profound effect of lignin degradation was shown in old barrels containing Armagnac for 20 years.[7] The inner and outer faces of the staves were cut (2 mm thickness) and the aromatic aldehydes content in the wood was measured. The results are shown in the following figure based on the above data:

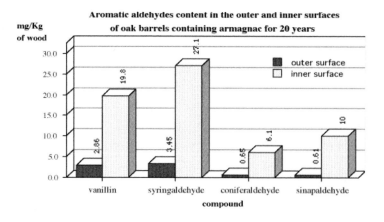

The data show that lignin in contact with the alcohol-acidic solution, is decomposed to its basic aldehyde components, which diffuse out from the wood into the liquid.

The concentrations of aromatic aldehydes during barrel aging were also measured with Cabernet Sauvignon wine, in two barrels sizes, 225 and 500 liters [8]. The results are shown in the following figure based on the above data:

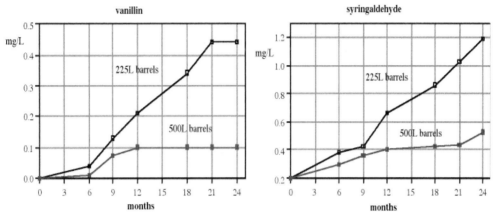

Vanillin and syringaldehyde concentrations in Cabernet-sauvignon wine aged in oak barrels of 225L and 5ooL volume size. The same wine aged in concrete tanks showed only traces of these aldehydes.

The effect of the smaller barrel size on the extraction rate is very clear.

Other identified aromatic *phenolic components* were found, such as:

(1) Eugenol which has a spicy clove-like flavor (and in fact is the main clove flavoring component), is extracted mainly from un-toasted oak. Its highest concentrations are found when the wood is just cut before seasoning. In seasoned wood its typical concentration is about 2–3 mg/Kg (a reduction from a typical value of 6 mg/Kg in green timber). The odor threshold of eugenol in white and red wines, are 0.18 and 0.70 mg/L respectively.

(2) Guaiacol and 4-methyl and 4-ethyl guaiacol are degradation products of thermal decomposition of lignin, and have a smoky flavor.

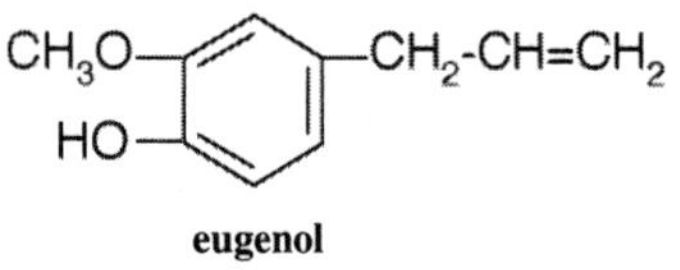

eugenol 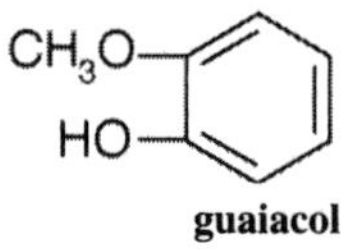   guaiacol 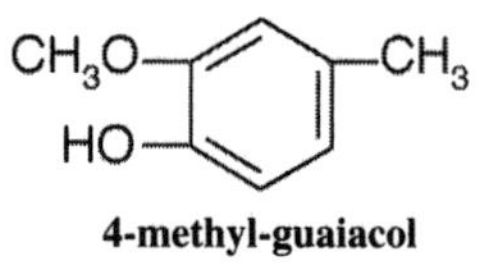   4-methyl-guaiacol

The odor thresholds of guaiacol in white and red wines, is 20 mg/L.

γ-lactones: Compounds with very significant contribution to oak flavor are the two isomers *cis* and *trans* *γ-octalactones* (4-methyl-5-butyldihydro-2-furanone, or simply b-methyl γ-octalactone). They were first identified in oak extract, and then recognized as important flavor compounds in whisky, brandy and wines aged in oak barrels. γ-Lactones are diastereoisomers (two asymmetric carbons), and hence there are four isomers, two in each of the *cis* and *trans* forms:

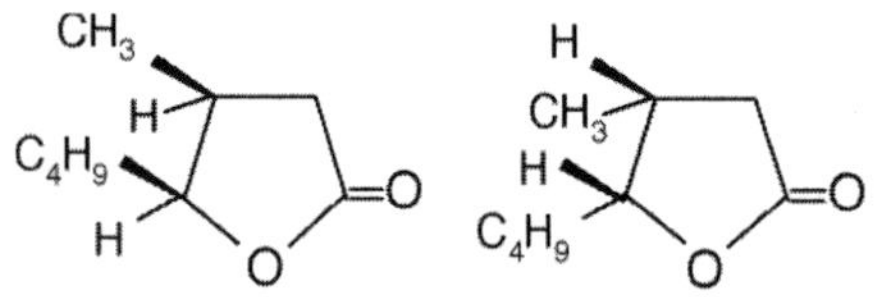

cis-b-methyl-γ-lacton 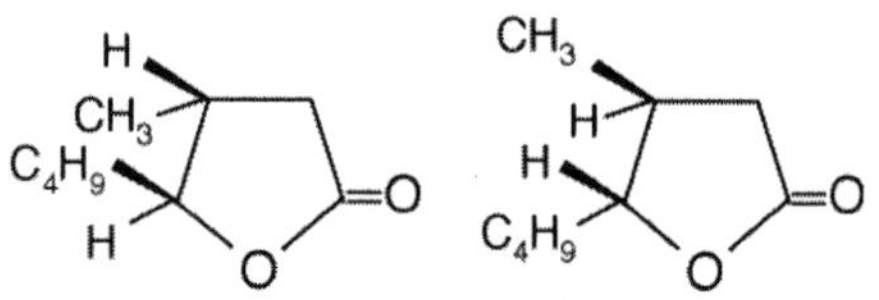   trans-b-methyl-γ-lacton

After identification of these lactones, there was some confusion in the literature regarding the correct isomeric assignment. For example, odor thresholds and other characteristics (GC retention time, mass spectra, concentrations in distilled beverages), of the two forms cis and trans [9], were given for the wrong isomer. The absolute configurations (shown here) were confirmed later.[10]

The odor thresholds of the two γ-lactones are quite different. The threshold of the *cis* isomers was found to be 0.067 mg/L, and the *trans* isomers are 0.79 mg/L (in 30% alcohol solution).[9]

The woody, coconut, oaky aroma of the lactones is therefore attributed mainly to the *cis* γ-lactones (ten-fold lower odor threshold). γ-lactone concentration in oak wood is in the 10–50 mg/kg range.

It is worthwhile to mention that there are also other lactone compounds which are found in wine, and which are not contributed from oak aging (mainly in sherry).

Other related aroma components of toasted oak: Which derived mainly from its hemicellulose degradation products, are *maltol, cyclotene* , and *ethoxylactone* :

ethoxylacton   cyclotene   maltol

White Wine 'Sur Lees': The effect of fermenting white wine in barrels and leaving it on its lees for few months (over a year in certain cases), was compared with the other alternative, such as tank fermentation and then racking and yeast removal before barrel aging for 11 months. In most extraction aspects, there was no significant difference between the two cases, in the composition of most compounds under study, including eugenol, guaiacol, 4-methylguaiacol and γ-lactones. However, there was one large difference. It was the vanillin content, which was much lower in the 'sur lees' wine. This was caused by the yeast biomass, which interacts with vanillin, leaving a very small amount of it in the wine. The interaction takes place during fermentation and afterwards. This interaction explains why barrel fermentation (and 'sur lees' aging) do not impact much of the oak character of the wine (although in most cases, new barrels are used). This effect was previously shown, where in added oak extract to fermenting model wine, the vanillin practically vanished. Or in practical winemaking, wines fermented and aged in barrels had lower content of vanillin (and vanillin flavor) than wines fermented in stainless steel tank and then aged in barrels for the same time.[11]

2. French vs. American Oak

The question of French vs. American barrels has some significant economic impact on wine production. Assuming three years of barrel usage (60 gallon barrel, which is equivalent to 300 bottles) and average storage time of one and a half years for each vintage (namely, 600 bottles per barrel lifetime during two vintages), the difference in oak cost of French over American barrels is about 0.7$ per bottle. It is not very significant in high priced wine, but it is significant in the fighting varietal segment of the market. So, what are the actual differences between the two-barrel categories?

The basic botanical difference is the oak species as mentioned above. The French barrels are made of *Quercus robur* and *Quercus sessillis*, where most of the American barrels are made of *Quercus alba* (about 50%), and other species such as *Q. bicolor, Q. macrocarpa, Q. prinus, Q. stellata,* and *Q. lyrata*. Other factors such as soil and climatic differences, growth rate, cutting location of the staves in the tree, and cooperage practice (wood drying, staves cut methods, fire or steam bending, and toasting methods and level), also contribute to the differences between the two categories of cooperage. Many studies had been done on this subject from various aspects.[12, 13] The results are controversial and unclear, and more has to be done to clear this matter.

Studies on oak extract, either as total solids or total phenols shows a remarkable difference between the two.[12b,12c, 1] French oak (or in general European oak) contributes more solid extracts and phenols than American oak. The results in one of the studies[12b] where the extraction was done with 55% alcoholic solution (which represent brandy aging in wood, but not wine), the total phenols and the nonflavonoid phenols extracted from 1 gram of wood (as chips) per 1L of solution are shown in the following figure:

Total and non-flavanoid phenolic extraction from 1 gram of French and American oak chips in 55% ethanolic solution

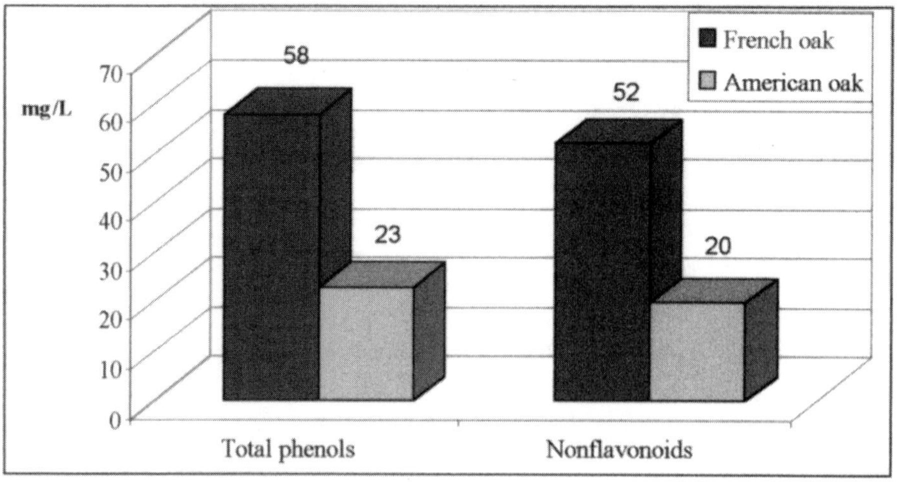

About half of the French oak chips (compared with American oak) were then considered to be necessary in order to give a treated wine a detectable threshold of oakiness (1.1 gr/L of French oak chips vs. 2.2 gr/L of American oak). This is not really the actual case because when specific nonflavonoid analysis was carried out (in a different study), although, gallic acid, protocatechuic acid, sinaptic acid, caffeic acid and total phenols were found to be higher in the French oak species (*Q. robur*)[13a] the concentration of vanillin was found to be higher in American oak (*Q. alba*). But the last compound has the highest impact on oakiness, compared to other oak components, and therefore American oak should actually give wine a more 'oaky' impression then the French one. On the other hand, as the total phenols (including tannins) are higher in French oak, the oak *"astringency"* is expected to be more noticeable in the French barrels (which is not the case because the extracted nonvolatile phenolics are below their taste threshold,[13e] see above in this section).

Specific measurements of the major *nonflavonoid tannin*, showed that higher extract was found in French oak, in comparison with American oak [12a]. Details on some other differences in basic components of wine stored in the two different oak categories were studied in another work.[13b] Chemical analysis of Cabernet Sauvignon wine aged in French and American barrels (same cooperage method i.e. air-dried, fire bending, and light toasting) revealed the following results:

(1) Phenolic extraction (total) from French barrels was measured to be only slightly higher than American ones, as can be seen in the following figure based on data from the above study[13b]:

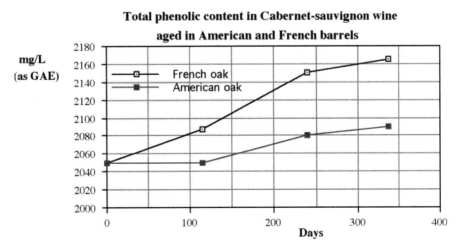

Total phenolic content in Cabernet-sauvignon wine aged in American and French barrels

In the above case, the accumulated phenolics extraction during almost one year from the American and French barrels was 2% and 5.5% of the total wine phenols, respectively.

(2) Titratable acidity (TA) increases during aging time, slightly more so when the wine is aged in French oak compared to aging in American oak. The rise of the TA concentration is attributed to evaporation from the barrels, and to acid extraction from the wood. The pH has a decreasing tendency over the same period (in both barrels), but it is not significant.

(3) Volatile acidity also increases over aging, with significantly higher concentrations in the French barrels. The increase in VA is related to volatile acids extraction from the wood. These results can be seen in the following figure based on the above study:

**Differences in TA and VA in wines stored
in French and American barrels**

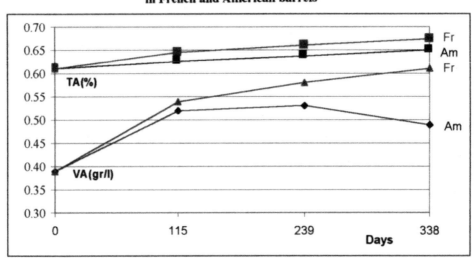

There is no comment on potassium bitartarate precipitation during the aging period, which may have some influence on the TA and the pH changes.

(4) Sensory difference evaluation of wine aged in French and American barrels.[13c] showed no significant difference between the two. In this case, maybe the sensory indifference is due to the heavy and highly phenolic content of the wine (Cabernet Sauvignon), and a general conclusion should be avoided. The difference between glass aged wine

and barrel aging (both in American and French) was very significant (p < 0.001). Of most interest is the finding that in both kinds of barrels, there was not any significant difference in wines that were aged for 115 days (4 months) in comparison to 338 days (11 months).[13b] Sensory descriptive analysis in model wines treated with oak chips of various origins revealed that American oak is perceived to be less intense in its spicy, nutty aroma then the French oaks.[13d]

3. Oak Alternatives

To shorten the time and expense in wine aging, experiments were done with oak chips or oak dust immersed in the wine. This 'fast-aging' method is not illegal, and if it is suitable to a specific wine, there is no reason why it should not be used. Even simulation of wine aging and brandy aging, by alcoholic (or water) oak extract, has been suggested and patented. The replacement of oak barrels by oak chips will partially fulfill only the extraction function of barrel aging, and not the oxidative one. To achieve the other function of barrel aging, controlled aeration has to take place, so let's look only at the extraction segment of the treatment.

The amount of nonflavonoids that can be extracted from oak chips (American) in red wine solution was measured to be 18 mg of total nonflavonoid as gallic acid equivalent (GAE) per 1 gram of chips per liter of wine.[12b] The amount of American oak chips needed, to be immersed in red wine in order to produce *oak recognition threshold* was estimated as 2.2 gr/L of wine. Or in other words, typical barrel-aged characteristics, for red wine, can be achieved with an American oak chip treatment of about 15 grams of oak chips per liter of wine. The extraction is quite fast and takes only a few hours to reach equilibrium. Sensory comparison between white wine aged in three different conditions, namely, new American barrels, used American barrels plus added oak chips, and stainless steel tanks plus oak chips, was carried on.[14] All the wines were aged to contain the same extracted quantity of gallic acid (as a probe for equal nonflavonoid extraction level). The sensory tests were very interesting. The aroma and flavor ratings of these different wine treatments, were practically equal. Also, the preferred quality rating was the same for all three wines, which raises some questions. Additional and very careful research is needed to study the question of barrel aging versus stainless steel plus oak chips. In practice, oak chips are sold in granular form as French and American oak. The chips are also available as light, medium and heavy toast chips.

Another alternative to the traditional oak barrel is innerstaves, which are installed in stainless steel tanks. This option is found in various shapes and tank sizes, but in principle the innerstaves replace the barrel as a container and function only as oak wood material for slow extraction. Also on the market are, 225L stainless steel drums containing interior oak staves held by a circular stainless steel holder. The stave can be replaced after one or two years of use. Another option is to replace the drum's heads with oak staves, which increases the wood area by 25%. Shaving old barrels has also been done for many years. Usually 5 year-old barrels (or more) can be shaved and reused for another two years with good results. The main problem that may arise after shaving is that the barrel's mechanical strength is weakened, so some staves may crack and leak, mainly at the bottom.

References

1. Singleton, V.L., "Using wooden cooperage in the winery today," *Vinifera Wine Growers J.* 8 (4) (1981), p. 227.
2. Technological report, "Cooperage," *Aust. & New Zealand Wine Ind. J.* 3 (1988), p. 33.
3. Singleton, V.L., "Wood and wine, the lasting marriage," *Am. Wine Soc. J.* (1984) fall issue, p. 71.
4. Lebrun, L., "French oak, origin, making, preparation and winemaking expectations," *Aust. Grapegrower & Winemaker,* Apr (1991), p. 143.
5. John, P., "Current cooperage practices," *Aust. & New Zealand Wine Ind. J.* 6 (1991), p. 66.
6. a. Graff, R.H., "Small cooperage: some practical aspects," *Wines & Vines 10* (1970), p. 27.
   b. Zimmermann, D., "Oak barrel preparation, care and storage," *Aust. & New Zealand Ind. J.* 6 (1991), p. 90.

c. Dodsridge, N., "Preparation, care and storage of oak barrels," *ibid,* p. 91.

d. Lienert, S., "Preparation, care and storage of oak barrels at Penfolds wines," *ibid,* p. 93.

e. Knox, M., "Barrel preparation and maintenance," *Practical Winery & Vineyard, May/June* (1987), p. 62.

7. Puech, Jean-Louis, "Extraction and evolution of lignin products in armagnac matured in oak" *Am. J. Enol. Vit. 32* (1981), p. 111.

8. Puech, Jean-louis, "Extraction of phenolic compounds from oak wood in model solutions and evolution of aromatic aldehydes in wines aged in oak barrels," *Am. J. Enol. Vit. 38* (1987), p. 236.

9. Otsuka, K. et al., "Presence and significance of two diastereoisomers of β-methyl-γ-octalactone in aged distilled liquids," *Agr. Biol. Chem. 38* (1974), p. 485.

10. a. Heide, R. et al., "Concentration and identification of trace constituents in alcoholic beverages," in A*nalysis of Food and Beverages,* ed. by Charalambour, G. (1978) p. 249.

    b. Masuda, M. and Nishimura, K., "Absolute configurations of quercus lactones from oak wood and chiroptical properties of monocyclic γ-lactones," *Chemistry Letters* (The Chem. Soc. of Japan), (1981), p. 1333.

11. a. Sefton, M. A. et al., "The influence of oak origin, seasoning and other industry practices on the sensory characteristics and composition of oak extracts and barrel aged white wines" – A*ust. Grapegrower & winemaker 355* July (1993), p. 17.

    b. Humphries, J.C. et al., "The influence of yeast fermentation on volatile oak extractives," *Aust. Grapegrower & Winemaker 343* (1992), p. 17.

    c. Chatonnet, P. et al., "Effect of fermentation and maturation in oak barrel on the composition and quality of white wines," *Aust. & New Zealand Wine Ind. J. 6* (1991), p. 73.

    d. Chatonnet, P. et al., "Effect of fermentation and maturation in oak barrels on the composition and quality of white wines," *Aust. & New Zealsnd Wine Ind. J. Feb* (1991), p. 73.

12. a. Quinn, M.K. and Singleton, V.L., "Isolation and identification of elagitannins from white oak wood and an estimation of their roles in wine," A*m. J. Enol. Vit. 36* (1985), p. 148.

    b. Singleton, V.L. et al., "An analysis of wine to indicate aging in wood or treatment with wood chips or tannic acid," *Am. J. Enol. Vit. 22* (1971), p. 161.

    c. Guymon, J.F. and Crowell, E.A., "Separation of vanillin, syringaldehyde and other aromatic compounds in extracts of French and American oak woods by brandy and aqueous alcohol solutions," *Qual. Plant. Mater. Veg. 16* (1968), p. 320.

    d. Guymon, J.F. and Crowell, E.A., "Brandy aging: some comparisons of American and French oak cooperage," *Wines & Vines 51* (1970), p. 23.

    e. Guymon, J.F. and Crowell, E.A., "GC separated brandy components derived from French and American oaks," *Am. J. Enol. Vit. 23* (1972), p. 114.

    f. Rous, C. and Alderson, B., "Phenolic extraction curves for white wine aged in French and American oak barrels," *Am. J. Enol. Vit. 34* (1983), p. 211.

13. a. Miller, D.P. et al., "The content of phenolic acids and aldehyde flavor components of white oak as affected by site and species," A*m. J. Enol. Vit. 43* (1992), p. 333.

    b. Aiken, J.W. and Noble, A.C., "Composition and sensory properties of Cabernet sauvignon wine aged in French versus American oak barrels," *Vitis 23* (1984), p. 27.

    c. Aiken, J.W. and Noble, A.C., "Comparison of the aromas of oak and glass-aged wines," *Am. J. Enol. Vit. 35* (1984), p. 196.

    d. Francis, I.L. et al., "A study by sensory descriptive analysis of the effect of oak origin, seasoning, and heating on the aromas of oak model wine extracts," *Am. J. Enol. Vit. 43* (1992), p. 23.

e. Pocock, K.F. et al. "Taste thresholds of phenolic extracts of French and American oakwoods: The influence of oak phenols on wine flavor," *Am. J. Enol. Vit. 45* (1994), p. 429.

14. Wilker, K.L. and Gallander, J.F., "Comparison of Seyval blanc wine aged in barrels and stainless steel tanks with oak chips," *Am. J. Enol. Vit. 39* (1988), p. 38.

CHAPTER VI

BOTTLING

CHAPTER VI: BOTTLING

A bottle of wine is the final product of the whole process of winemaking. The prospective customer, before buying a wine, will have an eye on the bottle, and part of his/her decision to buy a certain wine will depend on how it looks. Furthermore, when the bottle is opened for tasting, it must meet most of the customer's expectations. In addition, the wine is kept in the winery cellar or warehouse for a certain period of time, and then at the store and at the customer's home before it is opened and consumed. Care must be taken to ensure that the wine will stay alive, and age in the bottle without spoiling. All these factors must be taken into account when planning for the final operation in the winery: bottling. This chapter covers some important aspects of bottling, which are, in principle, the same at any bottling scale, from tiny amateur winemaking, up to full-scale winery operations. The only difference is in the equipment (hand operated, semi-automatic, or fully automatic machinery), which is determined by the quantity of wine to be bottled.

A. Wine Container

1. Bottles

Traditionally, wine bottles are made in three major shapes: the *Bordeaux* type, the *Burgundy* type, and the *Alsace* (also *'hock'*) type.

Bordeaux   **Burgundy**   **Alsace**

These names, of course, lead to the origin of these shapes. According to tradition, each varietal wine is held in a certain type of bottle. This means that most of the world's wines are bottled in one of the above three shapes, according to the regional tradition of the major French appellations. It would seem odd, for example, to find Cabernet Sauvignon in a Burgundy bottle or Chardonnay in Alsace or Bordeaux bottles. However, there are some exceptions to this general agreement. For example, a bottle of Cabernet Franc from the Loire is bottled in Burgundy bottle, according to the regional tradition.

The colors of the Bordeaux and Burgundy bottles are green or brown for red wines, and white or greenish-white for white wines. The Alsace bottle, which is suitable only for white wines, comes in two colors: green in the Moselle style, and brown in the Rhine style.

It is important to mention that light may cause some photochemical reactions in wines, which will accelerate wine aging. Red wines which containing high concentrations of anthocyanins and other phenolic compounds are more susceptible to these reactions. Therefore, wines that are to be aged for a longer amount of time should be kept in a colored bottle. This is the reason why red wines are traditionally bottled in deep green or brown (Italian style) bottles. Some white wines, specifically those that are intended to age in the bottle, may also be bottled in green bottles (Chardonnay, Riesling, and others). The types of bottles, and some representative varieties, are summarized in the following table:

Bottle type	Varietal wine	Bottle color
Bordeaux	Cabernet Sauvignon	green
	Cabernet Franc	green
	Merlot	green
	Nebbiolo	green/brown
	Sangiovese	green/brown
	Zinfandel	green
	Barbera	green/brown
	Sauvignon Blanc	white
	Semillon	white
	Muscadet	white
	Muscat	white
Burgundy	Pinot Noir	green
	Sirah/Shiraz	green
	Gammay	green
	Grenache	green
	Chardonnay	white/green
	Viognier	white/green
Alsace	Riesling	green/brown
	Gewurztraminer	green/brown
	Sylvaner	green/brown
	Muscat	green/brown

Varieties that are closely related to those cited in the table are traditionally bottled in the same type of bottle as the varietal they are related to. For example, wines in the Riesling family, such as Emerald Riesling (Riesling x Muscat) or Muller Thurgau (Riesling x Sylvaner) are bottled in Alsace bottles. Practically all white German wines are bottled in Alsace bottles, because most German white varieties are Muscat related. Cabernet Franc, Malbec, Petit Verdot, and Merlot, which are associated with the Bordeaux region, are bottled in Bordeaux bottles. The same argument holds for all Pinot related varieties (e.g. Pinot Noir, Pinot Gris, Pinot Blanc, Pinotage), which are bottled in Burgundy bottles.

As for generic wines, there is no agreement which bottle is suitable, but in many cases, the regional tradition dictates the bottle style. For example, Chianti wines are bottled in Bordeaux style (although, there is a unique 'Fiasco' shape bottle for the simple Tuscany wines).

The common bottle size, by worldwide agreement, is 750 milliliters. There are also half bottles (375 milliliter), magnums (1.5 L), double magnums (3.0 L), and Imperials (6.0 L).

The neck of all bottle types and sizes (except the Imperial) has standard dimensions in order to fit the standard cork size. At the opening of the bottle, the internal diameter is 18–19 mm, and at 50 mm down inside the bottle, it is 20–21 mm (see figure in the next section). So the inside of the bottle's neck is tapered upward to the top. This configuration of the neck holds the cork firmly in place.

2. Corks

In principle, for fresh fruity wines that are to be consumed young, there are cheaper alternatives, such as plastic corks, crown caps or aluminum screw caps. However, in the case of high-quality wines, sold at medium to high prices, one can afford to use the more expensive traditional corking. Pulling the cork from a bottle of wine is part of the ritual of wine drinking. It is as much a part of the style and appearance of the product as the right bottle, its color and the label. The winery's logo is also usually printed on the side of the cork. For many years of bottle aging, no material is known to be better than cork. However, after a long time, even the cork may deteriorate and fragment, causing the wine to leak out and allowing air to enter. Very expensive wines, which are aged by collectors for many years, must be opened and re-corked after 25–30 years. The wide use of cork as a wine stopper started only at the middle of the 17th century. It is related (truly or not) to a monk named Dom Perignon, who looked for a solution to secure his sparkling wine in a closed container. Since then, the innovation of cork became one of the most important contributions to the development of the wine industry.

General aspects

The English word *cork* probably comes from the Latin word *cortex,* which means bark. The cork is made of the bark of the oak species *Quercus suber* and *Quercus occidentalis,* which grow mainly around the Mediterranean Sea (Portugal, Spain, France, Italy, Morocco, Algeria and Tunisia).

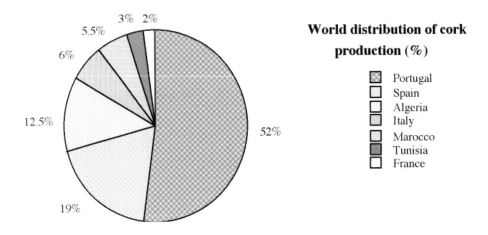

World distribution of cork production (%)

- Portugal
- Spain
- Algeria
- Italy
- Marocco
- Tunisia
- France

A new tree has to grow for about 20–25 years before the first bark can be peeled. It is then very dense and its structure is irregular, and therefore of poor quality and not suitable for use as a wine cork. The second bark is peeled off 8–12 years later. It is also of poor quality and therefore of little use. Only the bark of the third harvest is good enough to be used as a quality wine cork. That means that about 40–50 years are needed for a new oak tree to produce good quality corks. Nowadays, very few firms would start such a business. The time interval between each bark harvest is approximately 9–12 years, which is needed for developing suitable bark thickness. The number of annual rings can be seen on the flat cut of the cork. The useful life length of an oak tree as a cork source (from the time of the third bark removal) is about 100–150 years (although the tree may live for many centuries). General information on cork can be found in the listed books and papers.[1]

Cork production

The production of cork stoppers is a long process. The bark is peeled from the tree and allowed to dry and cure in the open air for several (six or more) months. It is then boiled for one to two hours to soften its texture, and to relive its internal stress, which changes the original bent pieces to flat bark bars. The boiling eliminates insects and other living objects from the bark, and also removes volatile phenols and other volatile substances. After a rest period of a few weeks, the bars are cut into strips whose width is the desired cork length. The strips are punched with a drill of the desired diameter to obtain the cylindrical corks. The raw corks are washed, bleached and vacuum-dried to suitable humidity (5%–8%). The corks are then graded by mechanical and/or manual screening to several cork grades based on faults and defects on its surface. After grading, the corks undergo surface treatment with various food-grade coating materials (paraffin, silicon oil, wax). The object of surface coating is to ease the insertion of the cork into the bottle, and also to ease its removal when the bottle is opened. The corks are then sealed in plastic bags containing SO_2 gas to ensure some sterility on shipment. As mention above, the corks are graded on a scale from poor quality to very high. Most of the cork producers grade their corks into 6–8 quality categories. The average percentage distribution of cork grades is shown in the following figure:

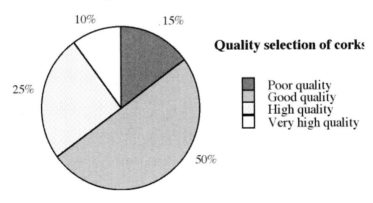

159

Physical properties of cork

The geometry of cork cells and the air content inside (about 60%–85% of its cellular volume is filled with air) reflects most of its mechanical and physical properties, such as low density, elasticity, compressibility, impermeability, adherence, and heat insulation. A cut through the three axes of the cork, namely radial, axial, and tangential (see figures), and a microscopic study, shows its cells structure.[2]

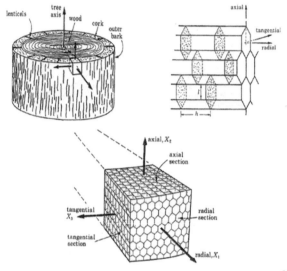

Diagram of the shape of cork cells in its three faces.[2]
The cells structure is an hexagonal prism whose base is in the radial section.

The radial face shows a hexagonal structure like a honeycomb, while the axial and the tangential faces are seen like a stack of bricks in a wall (see the microscopic figure).

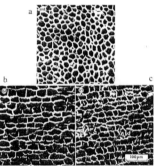

Electron microscope photograph of (top) radial, (left) axial, and (right) tangential sections of cork.[2]

The radial face is not always perfect and other polygons cross-sections are seen. Typical dimensions of the hexagonal prism cells are: Wall thickness—1μ; Prism edge length—20μ; Prism length—40μ; Number of cells per 1 cubic centimeter—approximately 10–40 million cells.

When making wine corks, one would expect the cork to be cut through the radial faces, to maintain the symmetry along the cork, which yields a better seal. Unfortunately, this is not the case, because of the *lenticels*, which are tiny channels connecting the outer and the inner sides of the bark. By cutting the cork in the radial direction, the wine will likely leak out. Also, by cutting the bark in the radial direction, there will be not enough bark width for the common 35–50 mm cork length. Therefore, the corks are cut from the bark in the axial/tangential direction.

A summary of some of the cork's properties is as follows:

(1) Low density: The density of cork is in the range of 0.12–0.20 gr/cc. A good quality cork has a density of 0.13–0.18 gr/cc. Therefore, the weight of a high-grade corks, of 24 mm diameter and 44–50 mm length will be in the range of 2.6–3.6 gram, and 3.0–4.0 gram, respectively. Higher density than 0.18 gr/cc indicates low elasticity and poorer quality.

(2) Elasticity or Resilience: Is a fast recovery from deformation made by compression. Under stress the air inside the cells is held compressed, and when the stress is removed, the pressurized air in the cells exerts force to restore the original shape. In cases where the cork is being stressed for a long period (as is the case when the cork is in the bottle), partial air content inside the cells diffuses slowly out, and when the stress is removed, the cork does not entirely regain its original shape and size. A bottle's neck diameter is about 18 mm, and most cork's diameter is 24 mm, which means a reduction of a bottled cork by about 25% of its radial dimension, or 45% of its volume (which is between points a and b in the following stress-strain curve). When the cork is pulled after being in the bottle for a long period, it does not regain its original size. In sparkling wine corks, where the original cork diameter was about 30 mm, the volume reduction in the bottle is almost 65%, and when it is pulled it gets the typical mushroom shape of sparkling wine corks.

(3) Compressibility: The ability to be compressed without permanent change of the original structure. A typical stress-strain curve of cork material is shown here:

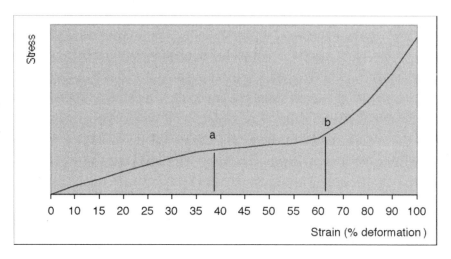

There is some linear section at low stress (up to point a). Where the cork's cell walls collapse because of the high stress, there is higher deformation at a lower extra stress (between points a and b). At very high deformation the stress-strain curve rises very sharply (above point b). At the linear section of the curve the deformation of the cork dimensions is still reversible. This is one of the most important characteristics of good quality cork. The cork should recover back to about 95% of its original shape after being released from compression, in a very short time (few seconds). In the actual process of wine corking the cork is squeezed and pushed into the bottle, and because of the smaller diameter of the bottle's neck the cork is not able to regain its original size. Therefore, the internal cork's pressure (within the cells) is not released, and thus the cork adheres firmly to the bottle's walls.

(4) Impermeability to liquids: Resistance to liquid penetration through the cork structure.

(5) Permeability to gases: Slow penetration of gases through the cork, by presumably microscopic channels ($\sim 0.05\mu;$) in the cells walls. At the diameter of such channels liquid cannot penetrate, but gases and vapors can slowly pass through. It was claimed that when the cork is compressed in the bottle, the condensed air inside the cells acts as a barrier to gas penetration. On the other hand, it is well known that after a certain time in contact with wine, corks gain weight by absorbing water vapors. Measurements on absorption of other vapors (phenol and chloroform) showed a remarkable increase in weight (about 30%). It appears that the subject of the permeability of gases through corks needs further investigation.

(6) Adherence: After being compressed and pushed into the bottleneck, cork has the property to strongly adhere to the glass, because of the desire to expand, caused by the squeezed cells.

(7) Insulation: A cork has excellent heat insulation, caused by the air held inside the cork structure.

(8) Immunity: Cork is very resistant to deterioration and rot decay.
All these properties make cork an excellent stopper for wine.

Cork taint

The phenomena of cork taint or "corkiness" as it is sometimes called, is characterized by a moldy/musty/smoky off-flavor. Cork-tainted wines must be distinguished from wines that have off-flavor caused by microbial fault, faulty barrels, or moldy grapes. There are two cases: one, where the faulty wines are discovered as individual bottles in a good sound batch, or, two, when the whole bottling batch is musty. The first case of individual musty bottles might be caused by faulty corks and is called 'cork-taint'. If the source of the fault is in the wine, then the whole batch will be off-flavored, and not randomly distributed. Estimation of the occurrence of cork-tainted wines during the last twenty years is about 2% of all wine bottles, which seems to be a serious problem.

The components found in corks, which might be responsible for cork taint, are:

2,4,6-Trichloroanisole (TCA) is a major component of wine fault caused by cork.

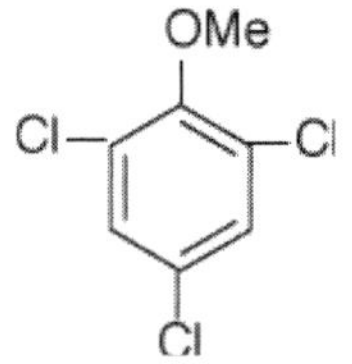

OMe

Cl—⬡—Cl

Cl

2,4,6-trichloroanisole
(TCA)

Its recognition as one of the major cork taints was determined in 1982, although it was known earlier as a food off-flavor component caused by fungal activity. Its musty/moldy off-flavor threshold is very extremely low. In water it is 3.10^{-2} ng/L, and in wine 4–10 ng/L.

TCA was also found to be responsible for the off-flavor in coffee, known as 'Rio flavor'. This compound was originally found in cork-tainted wines at a concentrations range of 20–370 ng/L, in comparison to less than 10 ng/L in untainted wines. In the above study, organoleptic tests made by addition of TCA to good wines at concentration range of 0 to 100 ng/L revealed that the musty off-flavor can be recognized at 10–30 ng/L. TCA is formed in cork in two steps. At first, phenols found in cork are chlorinated during the bleaching process with calcium hypochlorite to 2,4,6-trichlorophenol, which is then methoxylated to 2,4,6-trichloroanisole by fungal activity. To avoid such risk, some producers perform the bleaching with hydrogen peroxide rather than with hypochlorite. There are also other possible synthesis of TCA mentioned in the literature, such as the direct microbial synthesis from carbohydrates and chlorine source. TCA and other taint components (listed below) were also found in wines not being corked at all, but which were stored in oak barrels.[3] The source for these contaminations was attributed to microbial activity in the wood.

o-hydroxyanisole (guaiacol) causes wine taint which is described as a burnt, smoky, and medicinal off-flavor.

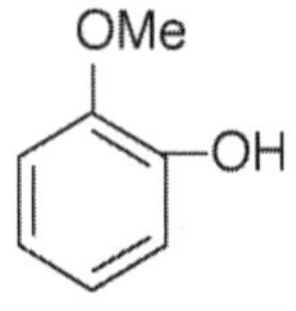

OMe

⬡—OH

2-methoxyphenol
(guaiacol)

164

Its off-flavor threshold in wine was found to be 20 µg/L, and in water 13 µg/L. In a study to find the cause of tainted wine, it was found that all wines described as tainted (9 samples) contained 70–2600 µg/L of guaiacol. In reference wines (3 samples of untainted wines of the same bottling) the concentrations in the wines were 3–6 µg/L. The traces of guaiacol found in sound wines looks to be reasonable, as this phenol was identified in trace quantities in corks, as a degradation product of lignin. It is also found in smoked meat and fish, roasted coffee and kiln dried malt as a phenolic pyrolysis product of wooden material (and also in toasted barrels) from lignin via ferulic acid to guaiacol.

1-octen-3-one and its corresponding alcohol **1-octen-3-ol** were also found in cork-tainted wines.

$$CH_3 - CH_2 - CH_2 - CH_2 - CH_2 - CO - CH = CH_2$$
1-octen-3-one

$$CH_3 - CH_2 - CH_2 - CH_2 - CH_2 - CH(OH) - CH = CH_2$$
1-octen-3-ol

The smell is described as mushroom-like, with the ketone odor threshold of 20ng/L, where the alcohol is 20µg/L (10^3 higher threshold value, much lower intensity, and therefore a very low contributor to tainted wines, if at all). The two diastereoisomers of the alcohol (carbon-3 is asymmetric) were identified (by NMR) and tested for their aroma characteristics and intensity. The R(-)-1-octen-3-ol was found to be responsible for the intense fruity mushroom-like flavor, while the S(+)-1-octen-3-ol has less intensive, grassy odor.

Trans-1,10-dimethyl-trans-9-decalol (geosmin) and 2-methylisoborneol

trans-1,10-dimethyl-trans-9-decalol
(geosmin)

2-methylisoborneol

are two other off-flavor components found in cork-tainted wines, with an earthy odor and a very low threshold of 25 and 30 ng/L, respectively. These compounds are natural byproducts of soil bacteria, which are also found in water supplies. The specified isomers (as written here) of these compounds are those that are responsible for their odor. Their occurrence in cork is probably due to contamination of the cork bark when it is cured in the open air, or later in storage.

To minimize cork taint, sterilization is essential. It was shown that fungi growth on cork contributes to "corkiness" flavor. Sterilization can be done by several treatments, such as sulfur dioxide, ultraviolet irradiation, γ-irradiation, and ethylene oxide. A summery of cork-taint compounds are given in the following table:

Compound	Off-flavor description	Threshold in wine
TCA	musty, moldy, wet cardboard	4–10 ng/L
Guaiacol	smoky, medicinal, phenolic	20 µg/L
1-octen-3-one	mushroom-like	20 ng/L
1-octen-3-ol	mushroom-like	20 µg/L
Geosmin	earthy	25 ng/L
2-methyl-isoborneol	earthy	30 ng/L

Extraction of volatile components from healthy corks reveals over one hundred compounds, which were identified qualitatively. It includes a wide range of chemical groups, such as hydrocarbons (aliphatic and aromatic), alcohols, aldehydes, ketones, esters, furans, sulfur and nitrogen compounds. Their contribution to the taste of wine is not yet known.

Practical aspects

The cork must fit into the wine bottle, according to its size and shape. A profile of the bottle's opening is shown in the figure:

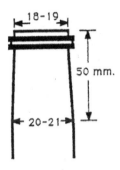

Wine's bottle opening (shape and dimensions)

The standard diameter of wine corks is 24–25 mm. The common length of wine corks is: 38 mm for simple and short term wines, 44 mm for medium-aging wines (the most common cork size), and 49–55 mm for long-aging and expensive wines. The dimensions of sparkling wine corks are 30 mm diameter, and about 50 mm length. Sparkling wine corks are generally made of granulated cork particles glued together (called the "body"), and one or two cork discs (about 5 mm width) at each side of the "body". With this structure the cork is capable of securing and holding high pressure (about 5–6 atmospheres) in the bottle. On corking, the cork is compressed by the machine's jaws from 30 mm to about 15–16 mm (half of its diameter), and then pushed into the bottleneck where it immediately tries to expand back to its original size. Because the inside neck diameter is between 18–21 mm, it remains compressed to about 60-70% of its original diameter, along the tapered bottleneck. In such a compressed state, the cork adheres tightly to the glass, preventing the sparkling wine from leaking.

After corking, theoretically, the cork may immediately regain about 95% of its original diameter, and a few hours to recover its full size. In practice, the maximum recovery is restricted by the bottleneck (about 75% of the cork diameter). So, the cork adherence to the neck of the bottle slowly increases with time after corking. The insertion of the cork into the bottle acts as a piston, and air is compressed into it. If the newly corked bottles are placed in the down position soon after bottling, the air pressure within the bottle (up to two-three atmosphere) and the partially recovered cork may cause the wine to leak. As a consequence, an upright position of at least several minutes, or better some hours, is recommended. Two methods can be applied in order to reduce the pressure in the bottle. The best is to cause a vacuum in the bottle headspace at corking, so that air is not compressed during corking. The second, and cheaper, method (when such a vacuum corking machine is not available) is to inject CO_2 into the headspace before the cork is pushed into the bottle. The dissolving of the compressed CO_2 in the wine will cause a considerable reduction of the pressure. This method has a disadvantage in that the dissolved CO_2 may be slightly felt in the wine.

One should also remember that if the wine is bottled cold (white wine), pressure will build up in the bottle when the wine reaches room temperature. The wine in a 750 ml bottle will expand about 2.5cc for an increase of 10°C. Therefore it should be remembered to leave at least 2 cm length of headspace (a volume of about 6.5 cc in the bottle neck). If, for example, only one cm of headspace is left (about 3.2cc) and the storage temperature will increase by 15°C, the *hydraulic pressure* of the expanded wine will push the cork out.

To test a new batch of corks for the presence of cork-taint components, a simple method has been suggested: choose randomly 20 corks and place in a closed beaker with one liter of neutral white wine which has not been in a wooden container. The corks should be fully submerged. After about 6–8 hours, smell the tested wine (with its reference wine) and look for a moldy, musty flavor. To cover the whole batch, the test should be done with a set of multiple samples.

Corks always contain dust, which may enter into the wine on bottling, and be seen mainly in white wines. Also, dried corks are difficult to slide through the corking jaws, and they tend to release more dust or small particles when bottled. It is therefore recommended to wash the corks for a very short time (minutes) with a solution of about 1000 ppm sulfur-dioxide, just before bottling. After rinsing, it is best to let them to dry for an hour. By doing so, most of the dust will be removed, and the wet corks will more easily slide into the bottle. If the washing is done for a longer time, and the corks are more saturated, when they are squeezed by the corking machine the corks release a brown extract containing cork-tannin into the bottle, which may affect the appearance of white wine.

Substitutes for natural cork

Good wine is always associated with a good cork stopper. But being a natural product, the variability of the qualities and performance of the corks has brought the need to look for other alternatives, which may be more uniform and predictable. Two such options are now in the market, namely, the *composite* and *synthetic* corks. The composite cork is made of agglomerated cork particles bound together with food-grade glue to form a uniform cork. In general, there are fewer leakage problems compared with natural corks, but the elasticity of composition corks is lower, which may create difficulties pulling them out. The synthetic corks are made of plastic materials (for example ethylene vinylacetate), and their surface is coated with silicone oil. The materials used comply with FDA food regulations. The corks are much more dense than natural corks (a typical cork weighs about 7–8 grams, compared to 3–4 grams for natural ones). Their permeability for liquids and gases is practically zero. The corks do not contain as many volatile compounds as natural corks, and they are not susceptible to the growth of microorganisms. Hence, no "cork taint" is expected to occur with synthetic corks. They are very uniform in structure and size; they do not break, dry-out or leak, contrary to natural corks. On corking, there is no dust problem with synthetic corks, and no washing is needed before use. Corking and uncorking is quite smooth with no breakage or fragmentation. The price is quite reasonable compared to natural corks. However, the use of synthetic corks is limited yet, mainly because of an image perception.

3. Capsules

Capsules are used to cover the cork surface and serve three functions: protecting the cork from cork borers, improving the bottle's appearance, and providing brand identity or authenticity. Traditionally, they have been made either of lead or wax. Nowadays, they are made of tin foil, aluminum or plastic. The plastic capsules are made either with thick walls and are fitted to the bottle's neck by machine, or with thin walls, which are shrunk under heat to fit tightly around the neck. The lead capsules were legally banned and were replaced by tin capsules, which have almost the same appearance, without any health risk. The capsules are fit onto the bottle's neck by mechanical shrinking. The winery's logo is usually printed on the capsule.

4. Label

The label on the bottle is the front window of the wine in the bottle. It must represent the winery and the wine in such a way that the prospective customer will choose this bottle of wine from the full selection he is offered in the shop. The label must show the following important information on the wine in the bottle:

(1) The kind of wine, namely, the variety of the grapes or its name in a generic wine.

(2) The name of the winery where the wine was produced.

(3) The geographical location of the vineyard, or in other word, its appellation (including the state, the county, and in certain cases the specific local village or specific vineyard). In each state there are specific rules concerning these issues.

(4) The vintage.

These four topics are the very basic details on the wine identity. In many cases, there is additional information on the label, such as:

(1) The classified name of the wine, which the winery selects to categorize the various quality selection of its production (e.g. basic quality, higher quality, and super quality).

(2) Regional quality status (e.g. Grand cru, Chianti Clasico, etc.) which is a general classification of the winery and not a specific quality description of the wine in that bottle.

(3) Also, by law in most countries, it should be clear from the label what the alcohol content is (v/v), as well as the wine volume.

(4) Other information, which the winery selects to have on the label (such as winery logo, *mis en bouteille au Chateau*, unfiltered, winemaker signature, and any other detail).

(5) Legal warnings. In the United States, as of January 1987, the label must note that the wine contains sulfur dioxide if there is more than 10 ppm of total SO_2. Also, other warnings in regard to health hazards have to be stated.

Some wineries put a back label on the bottle bearing additional information about the wine. This information is not formal, and usually includes some 'literature' information about the wine, the grapes, some details on the production process, the variety style, analytical data of the wine, barrel aging, some history, culinary/wine recommendations and more. In the US the legal warnings are usually placed on the back label.

The front label has to be approved by the authorities in most countries before the wine carrying that label can be released.

B. Bottling line

The final stage of wine production. It is important to remember that this last operation is irreversible—if some detail in the wine production was not done correctly, it will be impossible to correct it after bottling. The wine should be finished, stabilized in all aspects, filtered and sulfited, without any off-smell or taste.

All bottling materials must be ready (the right bottles, corks, capsules, labels, cardboards cases).

Flow chart of the line is shown below:

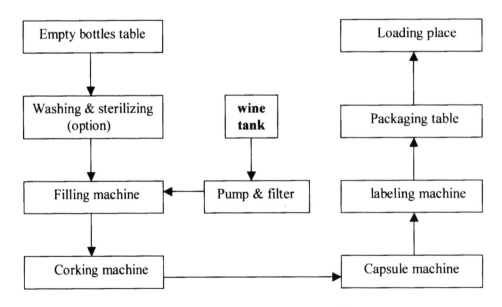

The bottling line assembly contains: an empty bottles table, a bottle washing and sterilizing device (optional), a pump and filter to transfer the wine from the tank into the filling machine, a filling machine, a corking machine, a capsule machine, a labeling machine, a packaging table to pack the bottles in cases, and a loading place for the wine cases. All the bottling elements must be connected with tables or, more conveniently, with conveyers, so the bottles can move easily on the line. In big wineries, the set is operated automatically, and only general supervision is needed to overcome occasional disorders in the bottling line. In smaller wineries, the system is generally not fully automatic and workers must be present to feed the various elements entering the line, such as bottles, corks, capsules, labels, and packing cardboards. In this case a minimum of five to six people are needed to run the line properly. The bottling machines that come in contact with the wine (bottling filter, filling machine, corking machine, pumps and hoses) must be clean and sterile.

Some useful points regarding bottling are mentioned here:

(1) Using the washing machine is optional. It is not necessary if the bottles were produced shortly before use (on coming from their production line they are clean and sterile). But if some months have passed, they may contain dust, insects and other materials, so it may be necessary to wash them before use. To avoid this, one should get the bottles a short time before planning the bottling, and insist on buying only freshly produced bottles. The washing device contains water jets and air jets to dry the bottles quickly after the washing.

(2) From the empty bottles table, the bottles are fed into the filling machine, which usually operates 6–12 filling stations in small machines, and up to 36–72 in a large one. When a bottle in the filling machine is full, various devices controlling the liquid level in the bottle halt the filling. The filling machine is fed with wine, which flows from the tank through membrane filter of 0.45μ in white wine filling, or some coarser filter in red one. In white wine bottling, it is recommended to place two membrane filters parallel, where only one is in operation, and the other is used if and when the pressure on the first filter become higher then 40 p.s.i. (above 2–3 atmospheres). When wine is flowing through the filter, the wines' particles are trapped on it, and the pressure may rise. The normal pressure difference across the membrane filters when clean water is flowing is about 0.5–1.0 atmospheres. Above 2–3 atmospheres, the second filter, without interrupting the bottling process, should replace the first filter. The first filter must be cleaned by flowing water in the reverse direction, and be prepared to be used as a spare.

(3) It is not recommended to use a pulse pump to filter the wine, because of the pulsed pressure being exerted on the filter. Good filtration should be done by a smooth and steady flow of the liquid.

(4) The next stage after filling is corking. This is done by a machine, which has "jaws," that contract the cork diameter before it is pushed into the bottle. The contraction should be done on the whole surface side of the cork with even pressure, otherwise the cork may be damaged and tiny chips of cork will be released into the bottle. The cork should be pushed down into the bottle exactly to the level of the bottle opening. After the cork is pushed into the bottle's neck, it tries to expand back to its original size. But the neck walls block the cork, and it is held firmly by the high friction constant of the cork surface and by the conical shape of the bottle's neck.

(5) The headspace between the wine and the cork should be about 1.0–1.5 cm. Not less, because of temperature expansion, and preferably not more because of excess oxygen in the headspace. Taking the thermal expansion of water as a base (0.0002 V/°C), 750 cc of wine in the bottle expand 0.15cc/°C. And assuming safe temperature increases of not more than 25°C, the wine will expand in the bottle about 3.7cc. The inner diameter in the bottle at the cork's end is 20–21 mm. Therefore, the volume of 1cm length below the cork, is about 3.2–3.5 cc. So, 1cm is the minimum length needed between the cork and the wine. On the other hand, a large space between the cork and the wine should be avoided in order not to allow much air to be trapped in the bottle. So, 1.0 to 1.5 cm of headspace is recommended.

(6) During corking a pressure may develop in the bottle (because the cork is pushed as a piston) which can rise up to a few atmospheres. It is preferred to

have a corking machine that exerts a vacuum in the bottle before pushing the cork in. In the case where the corking is not operated under vacuum, and if the bottles are placed head down (as is the usual way to pack and store wine cases), there is very good chance that the wine will leak out of the bottles unless the pressure in the bottle is released. It is therefore recommended in such a case to leave the bottles in the *up* position for a day to let the pressure to decrease, before placing it in the cases in the regular *down* position.

(7) The labeling machine is usually the "bottle neck" of the line. This machine causes most of the problems in the bottling line. Buying a good machine and reliable service is recommended.
It can apply three kinds of labels: the front label, the back label and the neck label. If one uses a neck label on bottles, they should know that this label causes the most difficulties in labeling, because of the tapered shape and small diameter of the neck.

(8) If there is any problem in the bottling line, and the bottling has to be stopped, the first machine in the line, namely, the filling machine, should be stopped, regardless of where the problem is. Also, after each station in the bottling line, it is recommended to leave some space for bottles to accumulate in case of some problem that may happen at the next station.

(9) The bottling machine and the membrane filter must be sterilized every working day before and after the filling. Passing hot water at 80°C for ten to fifteen minutes through the filter and the filling machine will do it. When starting the bottling, the first few bottles may be partially mixed with water and should be rejected from the line.

(10) If there is a need to touch the filling machine or the corking machine in order to fix some mechanical problem during bottling, it is recommended to spray the area with a 70% alcohol solution from a spray bottle after the problem has been fixed. The spraying can also be done from time to time without any specific reason, just to assure that those parts of the machines, which come in contact with the wine, are sterile.

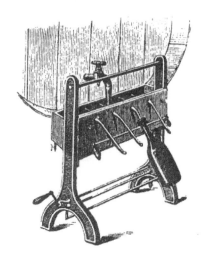

References

1.   a. Browning, B.L., *The Chemistry of Wood,* (1963).

     b. Sjostrom, E., *Wood Chemistry, Fundamentals and Applications,* (1981) p. 98.

     c. Faubell, A.L., *Cork and the American Cork Industry.*

     d. Technological report, "Cork, naturally the most suitable stopper," *Aust. & New Zealand Ind. J. 4,* Feb (1989) p. 31.

     e. Cooke, G.B., "Cork and its uses," *J. Chem. Educ.,* Aug (1931) p. 1462.

     f. Cooke, G.B., *Cork and the Cork Tree,* (1961).

     g. Casey, J.A., "Closures for wine bottles: a user's viewpoint," *Aust. Grapegrower & Winemaker,* Apr (1989) p. 99.

     h. Casey, J.A., "Cork as a closure material for wine," *Aust. Grapegrower & Winemaker,* Apr. (1989) p. 36.

2.   Gibson L.J. et al., "The structure and mechanics of cork," *Proc. Royal Soc. London A 377,* (1981) p. 99.

3.   a. Amon, J.M. et al., "A taint in wood-matured wine attributable to microbiological contamination of oak barrel," *Aust. & New Zealand Wine Ind. J. 2,* (1987) p. 35.

     b. Lee, T.H. and Simpson, R.F., "Cork Taints," *Practical Winery & Vineyard,* July/Aug (1991) p. 9.

CHAPTER VII

GENERAL ASPECTS

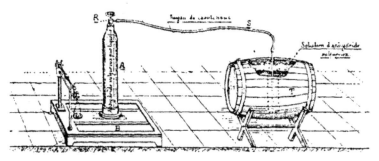

PRÉPARATION D'UNE SOLUTION D'ANHYDRIDE SULFUREUX.

CHAPTER VII: GENERAL ASPECTS

A. Sulfur-Dioxide

1. Sulfur Dioxide as a Food Product Preservative

Sulfur dioxide is a colorless gas, with a very strong, suffocating and irritating odor. Some of its physical properties are: molecular weight: 46.07; liquid form: at -10°C (one atmosphere pressure) and at 20°C (under pressure of 3.25 atm.); dissolves in water to form sulfurous acid (H_2SO_3). At 25°C the solubility is 8.5% (w/w).

Sulfur dioxide has been used as a preservative agent in the food industry for a very long time, probably since Roman times. In wine, sulfur-dioxide has been used from the Middle Ages up to the present. It looks almost impossible to produce wine without the addition of sulfur dioxide, for reasons we will discuss in this chapter.

Molecular sulfur-dioxide (SO_2), and sulfites (bisulfite HSO_3^- and sulfite $SO_3^=$) are multi action agents for the following functions:

- *Inhibition of nonenzymatic* oxidation of phenols, sugars and amino acids, which is followed by browning of plant tissues. The inhibition of this process is caused by the reaction of the sulfites with the carbonyl groups in the target molecule, blocking the chain of polymer formation.

- *Inhibition of enzymatic* oxidation, which also leads to browning products, for example in cut potatoes and apples, dried fruits, salad bars, fruit juices and grape must.

- *Antioxidant agent* in various systems, either by direct reaction with dissolved oxygen, or by reducing its oxidative products such as hydrogen-peroxide or quinones. In such reactions the sulfite ($SO_3^=$) is oxidized to sulfate ($SO_4^=$).

- *Inhibition of microorganisms growth* in food products, mainly bacteria, and in higher concentrations also molds and various species of yeasts. The exact mechanism of inhibition is not yet known, but the molecular state of sulfur-dioxide in water (SO_2), is the active form.

SO₂ Chemistry in Aqueous Solution

The equilibrium of sulfur-dioxide in water is (the pKa values are given at 25°C):

$$SO_2 + H_2O \Longleftrightarrow H^+ + HSO_3^- \; ; \; HSO_3^- \Longleftrightarrow H^+ + SO_3^= \quad (1)$$

$$pK_a = 1.81 \qquad\qquad\qquad pK_a = 7.2$$

At lower pH range where the first equilibrium is dominant, the dissociation constant K_d is: $K_d = [H^+][HSO_3^-]/[SO_2]$; where $[SO_2]$, $[HSO_3^-]$ and $[H^+]$, are the molecular SO_2, bisulfite anion, and proton concentrations respectively. And as pK_a is defined as $pK_a = -\log K_d$; and $pH = -\log[H^+]$; hence: $pK_a = -\log[H^+] - \log\{[HSO_3^-]/[SO_2]\} = pH - \log\{[HSO_3^-]/[SO_2]\}$; therefore:

$$\boxed{\log\{[HSO_3^-]/[SO_2]\} = pH - pK_a} \quad (2)$$

which means the ratio of the bisulfite form (HSO_3^-) to the molecular form $[SO_2]$ of sulfur-dioxide in solution is pH dependent. From this equation it is clear that when the pH of the solution is equal to the pK_a, then, in general, the ratio $\log[x]/[y] = 0$; $==>[x] = [y]$. Which means that the pK_a is an indication point, at what pH there are equal concentrations on both sides of the equilibrium equation.In the second equilibrium of equation (1) between the bisulfite and sulfite anions, the dissociation constant K_d is: $K_d = [H^+][SO_3^=] / [HSO_3^-]$ and the $pK_a = -\log[H^+] - \log\{[SO_3^=]/[HSO_3^-]\} = pH - \log\{[SO_3^=] / [HSO_3^-]\} = pH + \log\{[HSO_3^-]/[SO_3^=]\}$; therefore:

$$\boxed{\log\{[HSO_3^-][SO_3^=]\} = pK_a - pH} \quad (3)$$

The most important form of sulfur-dioxide as an antimicrobial agent is the molecular SO_2 form. In order to determine its actual concentration in solution, by measurable parameters, let us state the definition of "free" sulfur-dioxide. This term includes the concentrations of sulfur-dioxide forms, which are in equilibrium as shown above in equation (1) and are not bound to any compound in solution in an irreversible bond.

Free SO_2 is therefore the *sum* of $[SO_2] + [HSO_3^-] + [SO_3^=]$. Close examination of the sulfur-dioxide system in two pH ranges will reveal that:

a. In wine where the pH range is 2.9 to 4.0, the concentration of $[SO_3^=]$ is so small compared to the other forms (because of the pK_a value 7.2 which is very far from wine pH), that for all practical purposes, free sulfur-dioxide $\{[SO_2]_{free}\}$ can be defined as: $[SO_2]_{free} = [SO_2] + [HSO_3^-]$ which is the actual value that is analytically measured as free SO_2. By using the last expression of bisulfite in equation 2 which express the $[HSO_3^-] / [SO_2]$ ratio:

$\log\{([SO_2]_{free} - [SO_2]) / [SO_2]\} = pH - pK_a = pH - 1.81$

$\{[SO_2]_{free} - [SO_2]\} / [SO_2] = 10^{[pH - 1.81]}$

$[SO_2]_{free} / [SO_2] - 1 = 10^{[pH - 1.81]} ; ===> [SO_2]_{free} / [SO_2] = 1 + 10^{[pH - 1.81}$

therefore:
$$[SO_2] = [SO_2]_{free} / \{1 + 10^{[pH - 1.81]}\} \qquad (4)$$

b. At higher pH values where the second equilibrium is dominant (because of the same argument as before, low pK_a at high pH), the molecular form (SO_2) is so small that the free sulfur-dioxide can be written as $[SO_2]_{free} = [HSO_3^-] + [SO_3^=]$. And using this expression into equation (3) formulation for $[HSO_3^-] / [SO_3^=]$ ion ratio:

$\log\{[HSO_3^-]/[SO_3^=]\} = pK_a - pH ; \log\{([SO_2]_{free} - [SO_3^=])/[SO_3^=]\} = pK_a - pH = 7.2 - pH$

$([SO_2]_{free} - [SO_3^=]) / [SO_3^=] = 10^{[7.2 - pH]}$

$[SO_2]_{free} / [SO_3^=] - 1 = 10^{[7.2 - pH]} ; ===> [SO_2]_{free} / [SO_3^=] = 1 + 10^{[7.2 - pH]}$

therefore
$$[SO_3^=] = [SO_2]_{free} / \{1 + 10^{[7.2 - pH]}\} \qquad (5)$$

These two formulations enable us to calculate the important sulfur-dioxide agents, namely, the *molecular* SO_2 and the *sulfite* $(SO_3^=)$ concentrations at any pH of interest, by knowing *the free* SO_2 concentration. Or the other way, to estimate the free SO_2 needed to maintain certain value of molecular SO_2 (or sulfite) at a given pH. For example to maintain 0.8 ppm SO_2 at say pH = 3.5, the free SO_2 should be:

$[SO_2]_{free} = \{1 + 10^{[pH - 1.81]}\} \times [SO_2] = \{1 + 10^{[3.5 - 1.81]}\} \times [0.8] = \{1 + 10^{1.69}\} \times [0.8]$
= 40 ppm.

At higher pH, for example at say pH = 3.8, the free [SO_2] needed to have the same molecular SO_2 concentration (0.8 ppm) is about 80 ppm.

The distribution forms of the free sulfur-dioxide forms in the wine pH range is shown in the following three figures:

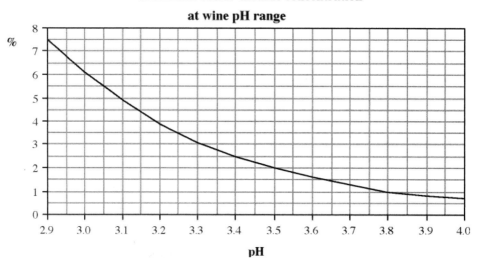

Molecular sulfur-dioxide concentration at wine pH range

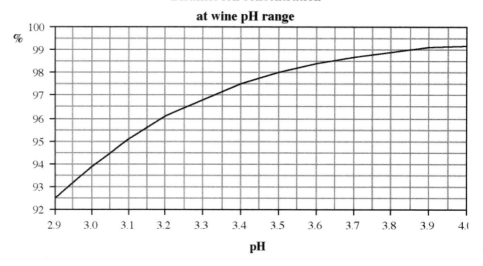

Bisulfite ion concentration at wine pH range

Sulfite ion concentration

at wine pH

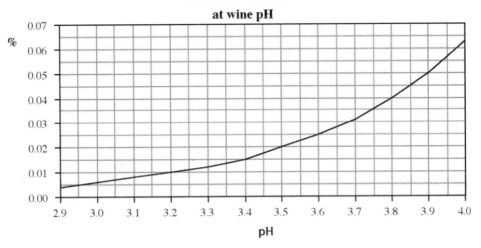

Most of the sulfur-dioxide (90%–99%) in this pH range exists as bisulfite ion (HSO_3^-), while the molecular form (SO_2) is at about (1%–7%). The sulfite ion ($SO_3^=$) exists only at a very low concentration range of 0.01%–0.1% range.

For the general interest, the full picture of the distribution of the three forms of sulfur-dioxide, namely, molecular (SO_2), bisulfite (HSO_3^-), and sulfite ($SO_3^=$), in a wider pH range, can be calculated by using the above formulae, and is shown in the following figure:

Distribution of sulfur-dioxide forms vs. wide pH range

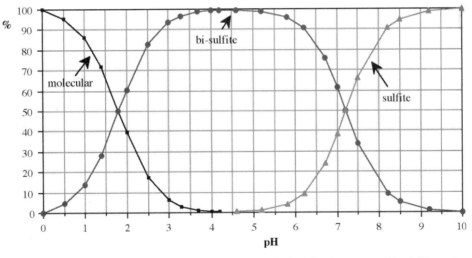

(Note the equal concentrations of molecular and bisulfite forms at pH = 1.81, and bisulfite and sulfite forms at pH = 7.2 as is expected by their appropriate pK_a. Also note the maximum percentage of bisulfite ion at pH 4–5, which is the pH boundary range between molecular and sulfite ion).

2. Sulfur Dioxide Uses in Wine

In this section we will discuss the specific aspects of sulfur dioxide use in wine production.

The Effect of SO₂ on Microbial Activity

The main yeast species active in wine fermentation are *Saccharomyces cerevisiae, Saccharomyces bayanus, Kloeckera apiculata*, and *Candida stellata*. At the beginning of 'natural' fermentation, the so-called 'wild yeast' species are mostly active. Later, the more alcohol tolerant species *Saccharomycess cerevisiae and Saccharomyces bayanus* take over to complete the fermentation. In 'cultured yeast' inoculation (based on the last two species) the specific yeast strain used for inoculation is practically the dominant one during fermentation, for two possible reasons: it is the most abundant in the culture, and because of the SO_2 addition to the must. In white must with up to 100 ppm of SO_2 added before fermentation, there is no practical effect on the *'cultured yeast'* growth, where at 50 to 100 ppm SO_2 addition, there is a delay of a few days in the growth of the *'wild yeast'*. (Addition of 150 ppm of SO_2 is needed to completely suppress the growth of these *'wild'* species)[1]. In red must, on the other hand, with addition of 100 ppm of SO_2, no difference was demonstrated between *inoculated* and *un-inoculated* fermentations. This is due to the presence of anthocyanins in red wines, which interact with the SO_2, reducing its inhibitory effect on yeast growth. In view of these findings and the practice of SO_2 addition during fermentation at about 20 to 70 ppm, and the use of 'culture yeast' strains, it looks that the control against 'wild yeasts' is not necessary and not effective. The effect of inhibition or delaying inoculated fermentation at these SO_2 addition levels, is insignificant.

The effect of SO_2 on the growth of *Saccharomycess cerevisiae* yeast species was studied [2] by experiments with radioactive SO_2. It was found that the molecular SO_2 is the sulfite form that is transported into the cells, a process that is completed a few minutes after application. Also it was found that the survival of the yeast cells after SO_2 application does not depend on the total SO_2, but on the molecular form. The molecular SO_2 concentration range depends of course on the pH, and for example (at pH = 3.19), a zero survival of the yeast cells after 5 minutes of application was achieved at 20 mmol of added SO_2 (about 1300 ppm), which is approximately a concentration of 50 (!) ppm of *molecular* SO_2 at this pH. (It should be remembered that the yeast contact with the SO_2 was for only a short time of 5 minutes).

In the practical lower levels of added SO_2 in winemaking, on inoculated fermentation with *Saccharomycess cerevisiae* yeast, there is some delay in starting the fermentation, and also in completion time, mainly at levels in addition of above 50 ppm, as can be seen in the following figure [3]:

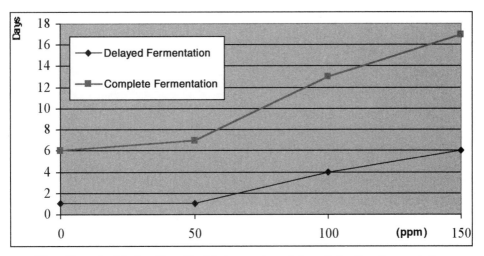

The effect of added sulfur-dioxide to must on delay of starting fermentation and time of completion (in S. cerevisiae at 21°C).

It seems that up to 50 ppm there is practically no effect on delaying and completing of the fermentation. Between 50 to 100 ppm which is still in the upper range of sulfite addition practice before fermentation, there is some delay (up to 4 days), and with another week to complete it.

On the other hand, there are some yeast species, which are very tolerant towards high SO_2 concentrations,[4] causing wine spoilage (e.g. *Saccharomyces bailii*), and others, which can maintain normal fermentation rate up to 300 ppm of added sulfur-dioxide.[3]

In secondary fermentation of sparkling wine, the free SO_2 level is very important with regard to problems that may arise in maintaining the fermentation. Levels of 25–30 ppm of free SO_2 are high enough to reduce very significantly the viable yeast population. The results may be a long delay in starting the fermentation, slow fermentation, and eventually stuck fermentation with low carbon-dioxide pressure. A safe free SO_2 level in a secondary fermentation is 10 ppm.

The preservation of grape juice from fermenting is of increasing interest, in cases where the juice is stored for later fermentation, or as a sweet additive to white wines. Two methods can be applied to preserve the juice: very high concentrations (1200–2000 ppm) of SO_2, where the juice is stored at ambient temperatures, and lower concentrations of SO_2 (200–300 ppm), where the juice is kept at low temperatures (0°C–5°C). In the first case, when the juice is needed for use, it has to be de-sulfited by heat and applying a nitrogen stream through it, leaving the juice with about only 100 ppm of SO_2. In the second case, fermentation can be delayed for a few months under these conditions. No mold or other spoilage bacteria (acetobacter, ML bacteria) grow under such conditions. Malolactic bacteria, which play an important part in winemaking practice (encouraging or preventing) is inhibited by the presence of SO_2, free and bound. The dependence of the ML fermentation rate on the total SO_2 concentration and pH values, can be seen in the Fermentation chapter section D.

Interactions of SO_2 with Wine Components

Only part of the total SO_2 concentration exists as *free* SO_2. The other part is bound to different components present in the must or wine. To demonstrate how much of the added sulfur-dioxide remains as free SO_2 in white *must* (Riesling) when it was added before fermentation, notice the following figure[3]:

The portion of sulfur-dioxide as free sulfite, out of the total added to Riesling must before fermentation (after 18 hours of addition)

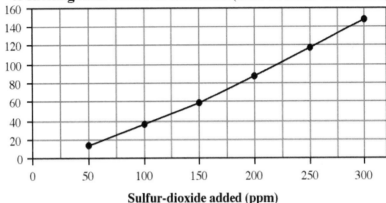

Free sulfite (ppm)

Sulfur-dioxide added (ppm)

Also, how much free SO_2 remains in a newly fermented white *wine* (Gewurztraminer) after addition of different quantities of sulfur-dioxide, where no addition was made before fermentation [5]:

The portion of sulfur-dioxide as free sulfite, out of the total added to newly fermented Gewurztraminer wine (after 5 days of addition)

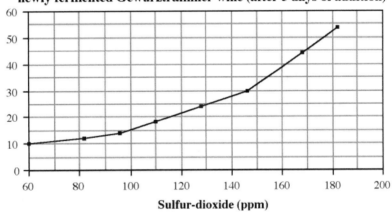

Free sulfite (ppm)

Sulfur-dioxide (ppm)

It looks quite clear that most of the added sulfur dioxide, especially at lower concentrations, does not remain as free SO_2. Hence it is bound and not available for its expected usage. The major components in must and wines, which interact with SO_2 to bind it, are carbonyl compounds, phenols, pigments, and some other specific components.

187

Methods of SO₂ Applications

As mentioned above, the antimicrobial activity of sulfur-dioxide to most of the wine related microorganisms, is the molecular form (SO_2). It was shown[6] that its effective concentration is at about 0.5 ppm–0.8 ppm. And because the distribution of sulfur-dioxide forms are pH dependent, the free sulfur-dioxide concentrations that are needed in order to maintain the desired molecular SO_2 concentration range, depend on the wine pH. This relation, is shown in the following figure:

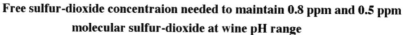

Free sulfur-dioxide concentraion needed to maintain 0.8 ppm and 0.5 ppm molecular sulfur-dioxide at wine pH range

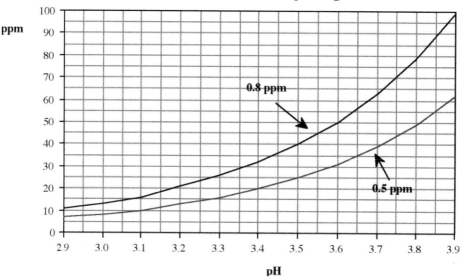

It is recommended therefore to store wines during cellar operations, with free sulfur-dioxide concentrations values, that are within the area between the two lines. For example, for a wine with pH = 3.5 the recommended range of free sulfur-dioxide concentration at bottling should be 25–40 ppm.

As was said before, when SO_2 is added to wine (or must), part of it reacts with some wine components to become bound and therefore inactive as an antimicrobial and an antioxidant agent. The rate of binding depends on wine state and age. It is very rapid when just added to must (a few minutes), and becomes slower as the wine gets older. The process of SO_2 becoming bound continues, which results in a constant decreasing of free SO_2 in wine at any stage of its life.

Additions of SO_2 are made at different stages of wine production. First at crushing, about 0 to 60 ppm is added to the must, depending on various considerations:

- Wine pH. The higher the pH, the greater the sulfite addition needed to be effective.

- Grape condition. The moldier the grapes, the higher the sulfite content bound by the mold substances. In such cases, the amount of SO_2 added should be twice that for healthy grapes.

- Malolactic fermentation. Because ML-bacteria are sensitive to SO_2, when malolactic fermentation is desired during fermentation, SO_2 addition should be minimal or not at all.

- When intentional oxidation and subsequent precipitation of phenolic components in white wines is desired, no addition of SO_2 is done at crushing.

After fermentation is over, a very small (if at all) concentration of free SO_2 is found in the wine, regardless of the initial addition, and practically most of the bound one (unless very high amount of SO_2 was added) is lost by oxidation to sulfate, and precipitation with the pomace particles and yeast cells. At first racking and afterwards during the whole process of winemaking, additions of SO_2 should be made in guidance of the above figure (0.5–0.8 ppm free SO_2). After fermentation, each addition is partially accumulated as total SO_2. Other part is lost (can not be detected by SO_2 analysis) to become sulfate ($SO_4^=$). A few days after any addition, analysis of SO_2 (free and total) should be made, with care being recommended so as not to exceed the total SO_2 above 150 – 200 ppm (although the legal limit in most countries is 350 ppm).

It should also be mentioned that even without any SO_2 addition, some sulfite is formed during the alcoholic fermentation by reduction of sulfate. The amount that is formed depends on the yeast strain, and it was found to be in the range of 10–30 ppm.

The most practical usage of sulfur-dioxide addition is by:

(1) Solid sodium or potassium meta-bisulfite which maintains the following equilibrium in water:

$$S_2O_5^= + H_2O \iff 2HSO_3^-$$

The HSO_3^- ion is in equilibrium with all other forms of sulfur-dioxide SO_2 and $SO_3^=$, and therefore an equilibrium is set that contain all three forms of sulfur-dioxide.

In potassium meta-bisulfite ($K_2S_2O_5$), only 57.5% of its molecular weight is SO_2, and therefore a very useful short formula may help to calculate the amount of potassium meta-bisulfite needed to add to a given volume of wine, in order to have the desired mg/L (ppm) of sulfur-dioxide:

| 17.5 gram of potassium meta-bisulfite/HL = 100 ppm | (6) |

(e.g. to add 40 ppm of sulfur-dioxide to tank full with 5000L, one should take 17.5x40x50/100 = 350 gram).

(2) Liquid SO_2 (density 1.4 gr/cc at 15°C) from steel tanks which are held at 3–4 atmospheres pressure. When the gas is released into the wine, it is practically almost totally dissolved as sulfurous acid, and the dosage can be monitored by weight lose of the steel tank or by a flow dosimeter.

B. Wine Spoilage

Being a food product, wine is susceptible to various chemical and microbial spoilages.

However, wine spoilage (in contrast to food spoilage) only causes faults in appearance, smell, and taste, rather than health-risking ones. The damage caused by wine spoilage is therefore only economic. The name of this section might also be *wine instability* (chemical and microbial), excluding *protein* and *potassium bitartarate* instabilities (which were discussed in the Cellar Operation chapter). The term instability is used to point out a process of undesirable compositional changes that occur in the bottled wine, which make its smell, taste and appearance undesirable or even unacceptable. Recognizing potential risks, preventing problems, and curing them if possible (once they happen), will be the subjects of this section. Only the most important and common faults will be mentioned here.

1. Chemical Faults

Metal Haze: This fault is quite rare these days. Although it still happens occasionally, it can be easily prevented. The metals of concern are iron, copper, and to lesser extent, aluminum. Although these metals are present in grape juice (in the range of 1–7 ppm of iron, and 0.1–0.4 ppm of copper), their concentrations are of such low values, that metal haze problems can be caused practically only by introducing them into the wine, for example, from equipment and accessories made of materials containing these metals. Contact of the acidic wine with such materials (iron, brass, bronze, and aluminum) dissolves some of them into the wine. The best way to avoid this problem is not to use any equipment or gadget containing these metals. The material of choice in the winery is stainless steel, plastic, wood, rubber, etc.

Iron: Iron haze is formed by ferric ion under oxidative wine conditions. In white wine the haze is composed of *ferric phosphate* (called "white casse"). In red wine it is composed of *ferric tannate* which is an iron-tannin complex (called "blue casse"). The maximal iron concentration where iron haze may *not* be formed is 5–10 mg/L. Above this level the potential risk is significant. To detect iron haze in wine, the cloudy wine sample should be acidified by about 1/4 of its volume with 10% HCl . If the haze disappears, *metal* haze is of significant probability. Addition of a few drops of potassium ferrocyanide (5%) to the acidified sample, which will cause a blue color, will indicate the presence of iron in the wine sample.

Copper: Copper haze is formed under reducing wine conditions, by interaction of the copper ion with proteins (in white wines). The maximum safe concentration of copper is about 0.5–1.0 mg/L. The protein content in wine also affects the potential haze formation. Besides equipment containing copper, another potential source of copper in wine is through the addition of copper sulfate to cure it from hydrogen-sulfide fault. Careful and exact quantities should be added, in order not to create a potential copper haze in the treated wine. To detect copper haze in wine, the cloudy wine sample should be acidified by about 1/4 of its volume with 10% HCl. If the haze disappears, it is most probably a *metal* haze. To verify it as copper haze, take a new sample of the cloudy wine and add a drop or two of 3%–10% H_2O_2. If the haze dissipates, the case is probably a copper haze. To complete the detection, a few drops of potassium-ferrocyanide (5%) should be added. Formation of a red color indicates the presence of copper.

Hydrogen-Sulfide: This is a very common fault in winemaking. Its typical smell resembles rotten egg. Most volatile sulfur compounds have a distinctive smell, with very low thresholds, and being a part (fortunately very small) of wine composition, they contribute significantly to the off-flavor and nose of wine. The sulfur for hydrogen-sulfide production may come from three sources:

(1) Yeast protein degradation, which releases hydrogen-sulfide.
(2) Residue of elemental sulfur dust left on the grape skin from antifungicide treatments in the vineyard.
(3) Reduction of inorganic sulfur compounds such as sulfate ($SO_4^=$) and SO_2.

The concentration of hydrogen-sulfide found in wine after fermentation ranges from trace amounts, to about 1000 mg/L, with a typical value of few hundred mg/L. Its off-odor threshold in wine and other beverages is quite low (values of about 50–100 mg/L). In air the detection threshold is about 20 mg/L. The more detailed forms from which hydrogen-sulfide can be formed are:

(1) Protein catabolism by proteolitic enzymes: Such protein degradation of sulfur containing amino acids (cysteine, methionine), may release hydrogen-sulfide. The major factor causing production of hydrogen-sulfide is *nitrogen* demand for the yeast growth. When this demand is not fully supplied, the nitrogen is gained by protein degradation processes, which also release hydrogen-sulfide as a byproduct from sulfur containing amino acids.

In order to control and limit hydrogen-sulfide production via this route, certain measures can be taken. Because the main source for hydrogen-sulfide formation is protein degradation, a decrease in protein content might lower its formation. But at the same time, a sufficient nitrogen supply, as free amino acids or as ammonia, will reduce the proteolitic activity needed to supply that demand, and therefore will reduce hydrogen-sulfide production. Adequate supply of nitrogen (as ammonium by diammonium phosphate (DAP) addition) prevents almost completely the proteolitic process during fermentation, and as a result controls and very effectively inhibits the production of hydrogen-sulfide. The following figure shows its effect[7]:

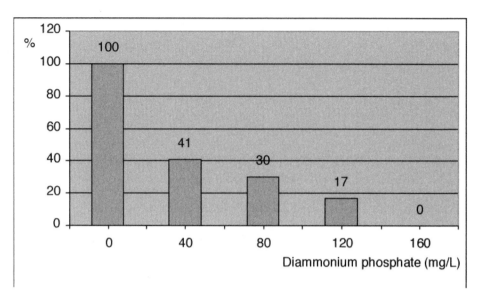

The effect of diammonium phosphate addition to must before fermentation on the formation of hydrogen-sulfide. The absolute value of 100% hydrogen-sulfide in this case is 422 mg/L.

The DAP addition is also effective even when added into the must *after* the hydrogen-sulfide is in the process of production. A few hours after addition, (while the fermentation is still active), the evolution of hydrogen-sulfide stops, and the bad smell disappears. The quantities of DAP that are needed to be effective, are (as can be seen in the above figure) in the range of 100–200 mg/L.

As a consequence, preventative addition of 100 – 200 mg/L of DAP prior to fermentation is a regular procedure in many wineries. Also, at the same time, the addition of a small quantity of bentonite to reduce the total amount of protein is also worthwhile.

(2) Elemental Sulfur: Sulfur is used to dust the vines against *powdery mildew* and other infections, and some of it remains on the grape skins. During fermentation and afterwards too, the elemental sulfur (S^0) may be reduced to sulfide ($S^=$). It is not clear if the reduction is an enzymatic or nonenzymatic process, although the last assumption is more probable. Experiments with adding elemental sulfur dust to a fermenting medium at a sulfur concentration of up to 100 mg/L, resulted in formation of up to 1200 mg/L of hydrogen-sulfide. The major factors that control hydrogen-sulfide production from elemental sulfur are: the temperature (lower production at lower temperatures), elemental sulfur concentration (higher hydrogen-sulfide concentration at higher sulfur concentration, with linear dependence at the range of up to 100 mg/L sulfur concentration), and ethanol concentration (higher hydrogen-sulfide formation at higher ethanol concentration, probably by increasing sulfur solubility).

(3) Inorganic sulfate and sulfite undergoing reduction to sulfide: Sulfate ions ($SO_4^=$) in the must, and sulfite added as sulfur-dioxide (SO_2), can be reduced enzymatically by the yeast cells to sulfides ($S^=$). This reduction is done in order to supply a sulfur source for the production of sulfur containing amino acids. When there is some shortage in the formation of these acids precursors, hydrogen-sulfide accumulates during fermentation.

There are also some other specific factors, which affect hydrogen-sulfide formation, such as *yeast strain, metals concentration*, and yeast *lees autolysis* :

(4) One of the characteristics of yeast strains is their tendency to form hydrogen-sulfide, which is typical to any strain. Some are high producers and some are low ones. Most 'wild yeasts' produce higher levels of hydrogen-sulfide. When choosing a yeast strain to carry on the fermentation, this characteristic may be of certain importance in a specific case. Usually the companies who produce yeast publish the characteristic details of each strain they produce, including their tendencies to form hydrogen-sulfide.

(5) Metal ions, when they are in excess, mainly copper (and iron too), stimulate the production of hydrogen-sulfide during fermentation. An earlier study did not find any metal effect on hydrogen-sulfide production. The question of how copper can stimulate hydrogen-sulfide production is speculated to be: either the yeast produces more hydrogen-sulfide in order to neutralize the excess copper by precipitating it

as copper sulfide, or by a natural selection of copper-resistant yeast mutate whose characteristic is to produce higher concentrations of hydrogen-sulfide. The copper concentration that is of concern in this matter is in the 1mg/L range and above. In this connection it is interesting to note that in some European wineries there is an old tradition of putting a copper bar in the must vessel containing the juice after crushing the grapes. This action is done to introduce some copper ions into the must, so any hydrogen-sulfide that might be formed during fermentation will react immediately with the copper ions, and precipitate as copper sulfide. This preventative action (to eliminate potential stink in the fermenting must), promotes more production of the agent that it is aimed to remove. It should be emphasized that only copper addition *before* fermentation may facilitate hydrogen-sulfide production. Addition of copper after fermentation is finished, in order to reduce excess of hydrogen-sulfide, has no such effect.

(6) When wine is left for yeast settling after fermentation is over for too long a time, the lees proteins may undergo autolysis, which produces hydrogen-sulfide. First racking a short time after fermentation is over (two–four weeks), might reduce hydrogen-sulfide accumulation from this source. The formation of hydrogen-sulfide may occur sometimes, even after the second racking.

Mercaptan and Disulfide: When hydrogen-sulfide is formed at high concentrations and is not removed from the wine in time, it may react with ethanol to form *ethyl mercaptan* (ethanethiol) CH_3-CH_2-SH, and *diethyl disulfide* CH_3-CH_2-S-S-CH_2-CH_3. The odor thresholds of ethyl mercaptan and diethyl disulfide are 1 and 4 µg/L respectively. These values are very low in comparison to other off-odor substances. Also, these thresholds are very small compared to their precursor, hydrogen-sulfide, which is about 50–100µg/L.

A simple test can verify the presence of a sulfide compound (hydrogen-sulfide, mercaptans, and disulfides) in a 'stinking' wine. The test is based on the following reactions:

a. Cadmium (Cd^{++}) ion interacts with hydrogen-sulfide, but *not* with the other two.

b. Copper (Cu^{++}) ion interacts with hydrogen-sulfide *and* with mercaptan, but *not* with disulfide.

c. Ascorbic acid can reduce disulfide to mercaptan.

To do the test, three glasses of the 'stinking' wine (about 20 ml) should be treated with the following additions (and mixed well):

- To the first glass – 1ml of cadmium sulfate solution (2 g/L).
- To the second glass – 1ml of copper sulfate solution (2 g/L).
- To the third glass – first, 1ml of ascorbic acid solution (10 g/L). After few minutes add to the same glass 1ml of the above copper sulfate solution. Now, smell the glasses, starting with a reference glass to which no addition has been made. Continue with glasses 1, 2 and 3.

The following table summarizes the seven possible results and findings:

Glass 1 (Cd^{++})	Glass 2 (Cu^{++})	Glass 3 (ascorbic + Cu^{++})	Findings
off-odor gone	off-odor gone	off-odor gone	hydrogen-sulfide
same as the reference	off-odor gone	off-odor gone	mercaptan
same as the reference	same as the reference	off-odor gone	disulfide
off-odor is < then refer	off-odor gone	off-odor gone	hydrogen sulfide + mercaptan
same as the reference	off-odor is < then 1	off-odor gone	mercaptan + disulfide
off-odor is < then refer	off-odor is < then 1	off-odor is < then 2 or gone	hydrogen sulfide + mercaptan + bisulfide
same as the reference	same as the reference	same as the reference	no detection of sulfides

If hydrogen-sulfide was produced during fermentation (in the case of nitrogen deficiency or due to other reasons), it should be removed from the wine as soon as possible. The simplest way to do it is by aeration (and some oxidation), namely, the whole volume of the wine is pumped from one tank to another one (or the same tank) while the wine is splashed over the top of the receiving tank. In most cases this measure is enough and the 'stinking' wine loses its off-odor right after this operation. In more severe cases, mainly when mercaptan or disulfide had been formed, chemical treatment is needed to remove it. To treat such wine, copper addition might help, up to 0.3 ppm of copper. The legal limit for copper addition is 0.5 ppm. Also, over-fining with copper may cause copper haze. Ascorbic acid addition (up to 25 ppm) followed by copper addition (few days later) may eliminate disulfide smell. If these additions reduce the bad smell, but not completely, carbon treatment (see Cellar Operation chapter, section C) might be of great help. In any case, a laboratory test to find the minimal amount needed is necessary before any addition is made.

Phenolic Haze: It is very rare although it may happen in white wines. Two types are known:

(1) **Flavonol haze:** Seen in recent years, in wines treated by the regular protein stability treatment, which then developed haze. It was shown to consist of the flavonols-protein complex. The specific flavonols in such haze formation (quercetin, kaempferol, myricetin) are found in quite low concentrations in grape juice. After being hydrolyzed (from their glycosides), they may interact with some proteins left over and form yellow sediment. The source for these flavonols was attributed to grape leaves found in the fermenting wine due to mechanical harvest, which introduces higher leaf content in the crush than hand harvest.

(2) **Ellagic-acid:** Which was observed in white wines treated with oak barrels or oak chips. A brown deposit was found in filtered bottled wines, a few weeks after bottling. The bottling was done in a short period after the wine was treated with fresh oak chips. The deposit was identified as *ellagic acid*, which was precipitated from the solution as a result of the hydrolysis of ellagitannins glycosides. Ellagic acid is insoluble in a wine medium. The ellagic acid precipitation was confirmed in model solutions containing 10% alcohol and saturated with potassium bitartarate. The solution was treated with oak shavings, then filtered and stored. A deposit occurred after a few days, which again was verified as ellagic acid. The phenomena of ellagic acid precipitation probably exists in red wine too, but was not noticed because of the

dark color and the normal expectation of a tannin deposit in red wines. In white wines aged in barrels, (or treated with oak chips), where the wine is bottled soon after being removed from the wood, such a brown deposit is a serious potential risk. It takes a few days to a few weeks for the deposit to be formed, so it may be advisable to delay the bottling for few weeks and then bottle the wine after membrane filtering. Then, the wine might be free of such problem.

Over Processing: This term includes operations in the course of the winemaking that have been done carelessly, or without attention to the impact on wine quality. In this term we can include such operations as filtering, fining, racking, sulfiting, and oak aging.

Over-filtering: Too much pad filtering, or with too much DE powder, may gives the wine a "papery" or "earthy" taste. Red wines should not be filtered during processing at all, except before bottling with coarse or fine filter, or unfiltered, as some wineries do. The best clarification of red wine which is aged is in barrels, is by natural precipitation. White wines can be filtered once or twice after cold stabilization, before sterile bottle filtration.

Over Fining: Fining with any agent (such as bentonite, gelatine, carbon, etc.) should be done with the minimum quantity required to fulfill its function, otherwise it will "strip" the wine of its aroma and flavor.

Over-racking: Will introduce too much oxygen to the wine, causing over-oxidation. Too much oxygen in white wines may show up as a yellowish color, lack of fruitiness, flat aroma, bitterness and even a caramel flavor. In rosé wines or light reds, the color will be a pink-brown or red-brown and usually accompanied by a flat aroma. Red wine pigments are oxidized and polymerized to lose their red color. They would age in the bottle quite rapidly, depending on the amount of tannins in the wine and their over-exposure to oxygen. In order to prevent this, minimum aeration should be practiced during wine processing, especially in white and "blush" wines. In general, red wines are less susceptible to oxygen, hence less vulnerable to racking operations.

Over-sulfiting: Sulfur-dioxide is added to the wine during its processing from crushing up to bottling. These additions are cumulative and at the end, the wine in the bottle will contain a certain quantity of bound SO_2 (which is difficult to recognize by taste), and a certain amount of free SO_2, which is described as a burning sulfur or pungent smell. The average threshold of free SO_2 is about 40–50 ppm. Excessive free SO_2 rarely happens in modern winemaking, as the SO_2 regime is strictly controlled. However, if it does happen, (probably by mistake), the way to reduce the extra free SO_2 from the wine is to strip it by *aeration* or by *oxidation*.

The aeration stripping can be done either by racking (from the bottom of the tank to the top of another one), or by bubbling nitrogen through the wine for a couple of minutes (from the bottom of the tank). The aeration method is not efficient enough in a wine with a high pH value. (Just to remind the reader, in the SO_2 analysis by the aeration-oxidation method, SO_2 is stripped from the analyzed wine by air bubbling through acidified solution of the wine).

The other option (oxidation) is more efficient and is based on the oxidation of SO_2 with hydrogen-peroxide: $SO_2 + H_2O_2 ===> SO_4^= + 2H^+$

To do it, follow this guideline: In order to reduce the free SO_2 in the wine by 10 ppm, add 50 ml of 1% hydrogen-peroxide to 1 HL of wine. An excess addition of peroxide may oxidize other components in the wine, reducing its quality. Therefore, the addition should be made only after a small-scale laboratory test has been performed to find the exact volume of peroxide needed to reduce the SO_2 to the desired level, without affecting the wine quality. The addition should be done in small steps, 10 ppm at a time.

Over-barrel aging: May cause the wine to taste and smell over-oaky which sometimes dominates the natural wine aroma and flavor. The wine is then unbalanced and its quality is definitely reduced. Tasting the wine during barrel aging is necessary to prevent such results.

Also careful selection of the barrels to match the wine at hand is recommended, depending on the barrel's condition and age.

2. Microbial Faults

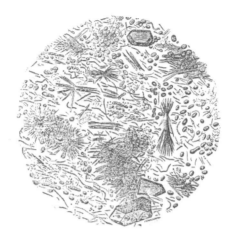

Microbial faults or instability are defined as the changes that occur in wine, as a result of microorganism growth, which cause chemical changes in the wine composition. The phenomena of wine spoilage are well known in wine production, probably since the time wine was first made. These changes may appear as turbidity, sediments, gassiness, color change, and sensory changes (off-odor and off-taste). Basically, wine is not a favorable medium for microorganism to grow. It contains alcohol at concentrations of about 10%–14%, its pH is quite low (3.0–4.0), and it contains various compounds that do not make life easy for microorganism to live in, such as sulfur-dioxide, higher alcohols (fusel oils), phenols, and others. In spite of these difficulties, some yeast and bacteria are able to find their specific substrate in wine and are capable of growing in it, if the right conditions are met. When such harmful microorganisms are able to grow in wine, we say that the wine is "sick". When the results of this sickness are noticed, it is usually too late to cure the damage to the wine (in the bottle and sometimes even before). For convenience, let us divide the various microorganisms that spoil wine, into two categories, namely, *aerobic* (which need air for their activity), and *anaerobic.*

Aerobic Spoilage
Acetic acid bacteria: This bacteria is very well known for changing wines to vinegar. The problem was so common in the winemaking industry that for example on the Davis score chart for wine quality evaluation, developed a few

decades ago, wine that was free of volatile acids (mainly acetic acid), was granted 10% of the total score, just because it was not spoiled with that fault. Such credit was abandoned some years later, because 'sick' wine containing high volatile acidity became quite rare. Acetic acid bacteria are classified into two genera, *Acetobacter* and *Gluconobacter*. Both genera make their living by oxidizing ethanol to acetic acid, but the difference between them is that Acetobacter are able to proceed further with the oxidation, to CO_2 and water, while Gluconobacter cannot. The Acetobacter genus, include four species, namely, *Acetobacter aceti, Acetobacter liquefaciens, Acetobacter pasteurianus*, and *Acetobacter hansenii*. The Gluconobacter genus includes one species *Gluconobacter oxydans*.

Several factors affect the survival and growth of Acetic acid bacteria in wine:

(1) **pH:** The optimum pH range for their growth is at about 5.0–6.5, which is much higher than the wine pH. The lower tolerance limit is at pH = 3.0 which is actually the lower pH range of wines.

(2) **Ethanol:** Although ethanol is their substrate, Acetic acid bacteria are sensitive to its concentration. Wine ethanol concentration (10%–14%) is quite high for their survival, although they were found to grow in wines of up to 15% ethanol, which is their upper-most limit.

(3) **Temperature:** The optimal growth temperature is about 25 °C–30°C. No data are available on the lower and upper temperatures that inhibit their growth, but from room temperature, their growth rate increases as the temperature is raised.

(4) **Oxygen:** Oxygen is vital for their growth. Any exposure of the wine to air may increase their activity, especially on the surface of stored wine in tanks or in barrels.

(5) **Sulfur-dioxide:** Has almost no effect on the Acetic acid bacteria growth in the regular concentrations of SO_2 in wine. Free sulfur-dioxide in the 25 ppm range is not enough to control their growth. More

information is needed in this matter. The above factors are interconnected in their effect on Acetic acid bacteria growth; for example, the alcohol tolerance is less at lower pH.

The over-all reaction that takes place by these bacteria is:

$$CH_3\text{-}CH_2\text{-}OH \xrightarrow{\;[O2]\;} CH_3\text{-}CHO \xrightarrow{\;[O2]\;} CH_3\text{-}COOH$$

The oxidation from ethanol to acetaldehyde and then to acetic acid is achieved by the use of atmospheric oxygen, which makes it a prime condition for the bacteria to proliferate. The acetic acid formed by these bacteria is added to the existing acetic acid and other volatile acids formed by the yeast activity. The legal limits of volatile acids (VA) in the US are 1.5 and 1.7 gr/L in white and red wines respectively. In a normal, healthy fermentation the VA concentration is in the 0.2–0.4 gr/L range.

The odor threshold levels of acetic acid in white and red wines are 1.1 and 0.8 gr/L respectively [11]. Wine containing acetic acid at concentration above 2–3 gr/L is unpleasant and is considered spoiled. *Ethyl acetate* (CH_3-CH_2-O-CO-CH_3) is always formed by the Acetic acid bacteria as a byproduct accompanying the formation of acetic acid. The sour odor of spoiled wine infected by the Acetic acid bacteria, is mainly caused by this ester rather then by the acid. Its odor threshold level (in white and red wines [11]) is in the range of 150–200 mg/L. When wine was infected by Acetic acid bacteria, beyond the spoilage level, there is no known treatment to cure it. The best way to avoid acetic acid bacterial growth, is never to leave any wine stored in a container that is not full. Or, if that is impossible, cover the top of the wine with a nitrogen or carbon dioxide blanket.

Flor sherry: Contains various genera of aerobic surface film yeast that grow on the wine surface when air is available, and consume its alcohol to produce acetaldehyde as their oxidation product. They form a white film on the wine surface, which are colonies of incomplete separation of the dividing cells after budding. The cells have a cylindrical shape with dimensions of about 2–4μ wide by 5–10μ long. These yeasts belong to the genera *Candida* (mostly

Candida mycoderma species), *Pichia* (mostly *Pichia membrafaciens* species), *Hansenula* (mostly *Hansenula anomala* species), and *Saccharomyces* (mostly *Saccharomyces beticus* and *Saccharomyces fermentati* species). In table winemaking (not sherry type wines) if these yeast infections are left on the wine surface for a certain time, they may spoil the wine with the undesirable acetaldehyde odor. On the other hand, they are encouraged to grow on sherry wines, where the acetaldehyde gives these wines their special and typical character. Certain conditions must be met for this yeast to grow. They are inhibited by a high level of alcohol (above 12%–13%), low pH, and sulfur-dioxide. In most cases the infection occurs in an un-topped tank. But sometimes the yeast grows, to some extent, in the bottle, as a result of non-sterile filtration, air exposure during bottling, or not enough free SO_2.

Anaerobic Spoilage

Fermentation Yeast: In wine that is not completely dry (containing more then 0.15%–0.20% fermentable sugar) when suitable measures have not been taken (sterile filtration, sorbate addition), re-fermentation will probably occur. It may happen during wine processing or in the bottle, where the damage is most harmful. When it happens in the bottle, it may cause some gassiness and sediment in mild cases, and bottle-explosion in the more severe cases.

Malolactic Bacteria: This is a very important microorganism factor in winemaking. More detailed information can be found in the Fermentation chapter, section D. Here it will be viewed from the perspective of potential spoilage. The wine is considered spoiled by these bacteria if they affect it in the bottle. The major reaction from these bacteria is the transformation of malic acid to lactic acid, followed by the release of carbon dioxide. Some other by-products such as diacetyl, acetoin, 2,3-butandiol, acetic acid and others are also produced. Some are desirable (buttery-like odor), and others are detrimental to the wine (acetic acid). Also, the bacteria are able to consume other metabolites present in the wine, which produce, in most cases, off-odor byproducts.

The malolactic bacteria are divided into three genera: *Leuconostoc, Pediococcus,* and *Lactobacillus.* The cultivated species used to induce malolactic fermentation is *Leuconostoc oenos,* which does its job rapidly and without undesirable off-odors. The other genera may cause some problems. *Pediococcus* is a slower fermenter, which leaves an unpleasant vegetative smell, mainly if the fermentation takes place in the bottle. The same happens with *Lactobacillus.* Lactic acid bacteria can also use other compounds as their substrate, like carbohydrates (glucose, fructose) even in a dry wine containing less then 0.5% of sugars. In fermenting wine, the risk is significant in cases where the wine was inoculated with malolactic bacteria, and the fermentation became stuck. The ML bacteria take the lead, using the sugar to produce mainly acetic acid. The VA of such a stuck fermenting wine becomes much higher than in normal fermentation. Immediate action to resume fermentation is vital to save the wine.

When sorbic acid has been added to preserve the wine from yeast fermentation (when some residual sugar is left over), malolactic bacteria may utilize it as a substrate to form *2-ethoxyhexa-3,5-diene* which has a very distinctive geranium-like smell. Lactic acid bacteria can also ferment tartaric acid, transforming it into lactic and acetic acids. The essential conditions for such sickness to happen are low acidity and low sulfur-dioxide. The results are reduction in fixed acidity, increase in volatile acidity, cloudiness, gassiness (carbon dioxide), and off-odors. In the old French literature this 'sickness' of wine was called "vin tourne".

To avoid the risk of malolactic fermentation in the bottle, it should be either encouraged to take place under controlled conditions during wine processing in the cellar (mainly in red and white wines that will age), or prevented by taking certain measures that are detailed in the Fermentation chapter (low pH, SO_2, inhibitors such as fumaric acid, and sterile filtration).

Brettanomyces/Dekkera: The double name of this yeast genus refers to different forms of the same microorganism. The asexual form (budding reproduction) is called *Brettanomyces*, while the sexual (sporulating) form is called *Dekkera.* This yeast is found in wines and beers. Not enough is known about this yeast, namely its metabolic and growing conditions. It can grow in certain wine, and not in others, with no clear understanding as to why. Red wines are its favorite. A low SO_2 concentration does not inhibit its growth, nor does sorbate. About 25 ppm of free SO_2 is needed for temporary inhibition.

When the free SO$_2$ becomes bound and reduced to a low level, it may resume its activity. Probably the most effective method to prevent its growth is sterile filtration. This is not always practical, at least in cases where red wines are not filtered (or just coarsely filtered) in order to minimize wine stripping. The distribution and infection of this yeast appears to be local in certain wineries, and it is transmitted within the winery through the equipment, barrels and tanks.

The organoleptic effects of Brettanomyces infection on wines start as a kind of "complexity" nose, but as the infection advances, the aroma is described as "mousy", "horse blanket", or "wet dog", with a metallic aftertaste. The "mousy" odor was recently related to the formation of two compounds *2-acetyl-1,4,5,6-tetrahydropyridine* and its imino tautomer *2-acetyl-2,4,5,6-tetrahydropyridine:*

2-acetyl-1,4,5,6-tetrahydropyridine   2-acetyl-2,4,5,6-tetrahydropyridine

These compounds, which have a characteristic "mousy" smell, were identified in "mousy" wines infected by Brettanomyces yeast and also Lactic acid bacteria. *2-acetyl-1,4,5,6-tetrahydropyridine* is also known to be a component in baked bread, with an odor threshold of 1.6 ppb. It was shown that the biosynthesis of these compounds is involved with the amino acid lysine and ethanol as precursors.

Saccharomyces bailli: This is quite unique yeast that can grow in very high sugar concentrations (e.g. grape juice concentrate), which means they can tolerate high osmotic pressure. They are highly resistant to sulfur-dioxide, sorbic acid and even to diethylpoly carbonate (DEPC). The yeast shows good resistance to all three agents at concentration range of up to 200 ppm. Being so unaffected by various inhibition agents, the only measure that can preserve an off-dry wine from being spoiled in the bottle by this yeast, is a reliable sterile filtration.

Some reviews on wine chemical and microbiological stability can be found in reference.[8]

C. Legal Regulations

Legal aspects of wine encompass a very wide range of activities such as establishing a winery, production, trade, taxes, and others. In every country that produces wine there are wine laws and regulations, which are tailored to the specific characteristics and needs of the people located in that country. Most regulations are supposed to protect the customer against risking his health and his pocket, and in return, to collect money (by taxes) back, from that pocket. One should not forget also that many administrative jobs are produced to maintain and supervise the wine regulations.

Like any other laws, wine laws reflect the attitude and feeling of the society to wine and wine culture. By focusing on one country, a good example of the concept, and an introduction to this subject can be gained. Technical regulations such as production procedures, wine content, and wine additives are more or less common to most countries.

Basically, wine regulations cover three phases of winemaking, namely, viticultural practice, wine production and wine trading. Viticultural laws regulate the appellation designation, the varieties, crops yield, and grapes trade. Wine production laws control the issues concerning public health and fraud prevention. The trade laws monitor the marketing system and the tax collection. Here we shall view only the production aspects of wine, and we shall do so through the regulatory system in the United States. The technical regulations are basically similar to any other country. In the US the federal wine laws are under the jurisdiction of the Department of the Treasury, with control of the subunit of the Bureau of Alcohol, Tobacco, and Firearms (?!), abbreviated BATF. This office is in charge of the production of wine, and collection of money (as established by the Internal Revenue Code). The Federal Alcohol Administration Act (FAA) controls the distribution, labeling and advertising of wine. Besides federal regulations, each state has its own control (and taxes) on wine production, appellation, and sale.

1. Definitions

Grape wine: wine produced by the normal alcoholic fermentation of the juice of sound, ripe grapes. Minimum alcohol content is not less then 7% (v/v).

Fruit wine: wine produced by the normal alcoholic fermentation of the juice of sound, ripe fruit. Minimum alcohol content is not less then 7% (v/v). If the wine is made from one kind of fruit it should be called wine with the fruit name (e.g. *berry wine*). If the wine is made from apple it may be called *cider.* If the wine is made from more than one kind of fruit it should be called *fruit wine.*

Agricultural wine: wine produced by the normal alcoholic fermentation of agricultural products other then grapes and fruits (e.g. honey, but not grain or molasses).

Table wine: grape wine, with alcohol content not in excess of 14% (v/v).

Dessert wine: grape wine with alcohol content between 14% and 24%. The alcohol added to fortify the desert wines should be grape alcohol (brandy). Specific wines in this category are *sherry,* which should contain not less then 17% alcohol. *Port* wine should contain not less then 18% alcohol. In a case where the alcohol content of sherry is between 14%–17%, and port between 14%–18%, these wines should be called *light sherry* or *light port.*

Sparkling wine: an effervescent wine containing more than 3.92 gr/L of carbon dioxide resulting solely from the secondary fermentation of the wine within a closed container.

Champagne method: is defined as sparkling wine which derives its effervescence solely from secondary fermentation within containers of not greater than one gallon, which also possesses the taste, aroma and characteristics attributed to Champagne wines. In most countries other than the US, such protected names (as Champagne) are not allowed. Sparkling wines made in larger containers (mostly in pressure tanks) cannot use the term Champagne method, but rather the term *Charmat method.* If the wine is charged by artificial carbon dioxide, it is called *carbonated wine.*

Appellation: has two definitions: *American wines*, and *American viticultural areas*. The first term specifies the state and county where the grapes are grown. The second term denotes specific viticultural areas, which have been approved as such. This appellation term does not necessarily overlap with the geopolitical boundaries definition, rather it is defined as a specific grape-growing region approved by the BATF. Sonoma County in California, for example, consists of eleven viticultural areas. Unlike other countries, the US appellation system does not indicate grape variety or viticultural methods control, and no quality criteria or characteristics of the wines made in the viticultural areas are required. The federal regulations require that at least 75% of the grapes must be from the state and county printed on the label. In California 100% of the grapes should come from California in order to use the state's name (California wine) on the label. For *viticultural areas* within California, at least 85% of the grapes should be from that area.

Geographical semi-generic names: are names that indicate specific characteristics of the wine and not their geographical origin. For example, Champagne, Port, Sherry, Chablis and others, are considered and recognized by their type of wine and not as having been imported from abroad.

Estate bottled: may be used on the label only if the grapes were grown within a viticultural area or appellation, and several other conditions are met: The bottling winery must be located on the land where the grapes are grown, all the wine must be made from grapes grown on this land, and the whole process of making the wine, including crushing, fermenting, aging and bottling must be done at that winery. No other term, which may indicate any connection between winemaking and growing the grapes, are allowed.

Varietal wine: the wine should contain at least 75% of the grape variety whose name is printed on the label. A blend of wine that contains less then 75% of one variety should have a generic name. Of course, any varietal wine (by the above definition) can also be called by a generic name.

Generic wine: wine made without one dominant grape variety making up at least 75% of its content. Such wine cannot be called a varietal wine.

Vintage wine: must contain at least 95% of grapes harvested in the year identified on the label.

2. General Regulations

Wine label: shall not be false or misleading. It should contain the following information:

(1) The name of the producer, or brand name.

(2) The bottler name and address.

(3) The kind of wine (e.g. varietal or generic table wine, desert wine, etc.).

(4) The vintage.

(5) The alcohol content (v/v), except in table wines where the alternative option is to write the words "table wine" which automatically indicates the alcohol content to be in the 7%–14% range.

(6) The bottle content in metric sizes. The most common bottle size is 750 ml in almost any country in the world. Other sizes are multiplications of the basic 750 ml, namely, 375 ml, 1.5 L, 3.0 L, 6.0 L and 12 L. Due to lack of uniformity in bottling, the volume of wine in the bottle shall not be *lower* by 1.5% in 1.5 L bottle (22.5 ml), 2% in 750 ml bottle (15 ml), and 3% in 375 ml bottle (11 ml).

(7) Shall contain certain warnings in regard to health risks such as alcohol use during pregnancy, sulfite content (if contain more then 10 ppm), and also the warning that alcohol may impair the ability to drive a car or operate machinery.

The wine label should be approved by the BATF.

Taxes: are paid according to federal *and* state regulations, by the wine producer at the time of removal of the wine from the winery. The taxes are progressively increased (based on alcohol content) according to the following categories:

(1) Table wine containing not more than 14% alcohol.

(2) Wines containing more than 14% alcohol and less than 21%.

(3) Wines containing more than 21% alcohol and less than 24%.

(4) Artificially carbonated sparkling wines.

(5) Sparkling wines.

When the product contains more than 24% alcohol, it is taxed in the higher category of *distilled spirit*.

3. Wine Content

The materials mentioned in this section are those controlled during wine production, which are considered *wine components*. The wine *treatment additives* are not included here. Wine treatment additives are presented in the next section.

Water: may be added to the must before fermentation, provided the sugar content of the must is not reduced below 22 Brix. This practice is not permitted in California. Instead, the unclear and ambiguous regulation in California states, that, "no water in excess of the minimum amount necessary to facilitate normal fermentation may be used in the production or cellar treatment of any grape wine". The practical meaning is <u>not</u> that it is allowed to use water regularly in wine production. Instead, it means, that in a case where the sugar content of the grapes is very high (e.g. above 24.5°B), water is allowed to be added, in order to lower the potential alcohol content to be below the legal value of 14%. In such a case, water addition should be done *before* fermentation.

Sugar: in the case of low sugar content, it is legal to add dry sugar or concentrated juice before or during fermentation. The sugar addition (*chaptalization*) should not raise the juice density above 25 Brix. Addition of 17–19 gram of cane sugar (sucrose) per 1L of must, will increase the alcohol content by 1%. This formula can serve as a general guide for the quantity of sugar needed. There is no definition of what is considered to be low sugar content, so in practice, sugar addition is permitted in the US. In California, no sugar other than pure grape concentrate may be used (before or after fermentation), except in the production of sparkling wines before the secondary fermentation (dosage).

Acids: tartaric or/and malic acids can be added to must before fermentation to correct acid deficiency. After fermentation, other acids such as citric, fumaric, lactic (beside tartaric and malic) can also be added up to a fixed acidity limit (calculated as tartaric) of 9 gr/L. There are specific limitations on the content of each specific acid in wine as follows: citric acid – not more than 0.7 gr/L and fumaric acid – 2.4 gr/L.

Volatile acids: mainly acetic acid. The current limits are 1.7 gr/L and 1.5 gr/L in red and white wines respectively.

Hydrogen-sulfide: there is no limitation on the content of this component in wine.

Methanol: shall not exceed more than 1.0 gr/L.

4. Wine treatment additives

A summery of the most useful materials used in wine production is given in the following tables. (The word GRAS in the tables mean Generally Recognized As Safe by the World Health Organization – WHO):

Fining materials

Material	Used for	Limitations
Bentonite (aluminosilicate)	Protein stabilization	GRAS. At 5% water slurry, not to exceed 1% water of the wine volume.
Carbon (activated)	Fining agent for removal of color and off-odor	GRAS. Up to 3 gr/L.
Casein	To reduce tannin content	GRAS.
Copper sulfate	To remove hydrogen-sulfide	Not to exceed 0.5 mg/L as copper, with residual level not to exceed 0.2 mg/L.
Egg-white (albumin)	To reduce tannin content	GRAS.
Gelatin	To reduce tannin content	GRAS.
Ferrocyanide (cufex)	To remove trace elements, and sulfides & mercaptans	GRAS. No residual level in finished wine in excess of 1 mg/L.
Isinglass	To reduce tannin content	GRAS.
Milk (pasteurized)	To reduce tannin content	Not to exceed 0.2% v/v of the wine.
Oak chips	To add oak flavor	GRAS.
PVPP (Polyvinylpolypyrrolidone)	To reduce tannin content	Not to exceed 7.2 gr/L. Should be removed by filtration.
Potassium bitartarate	Cold stabilization agent	GRAS. Not to exceed 4.2 gr/L.
Silica gel	Fining agent in combination with gelatin to reduce tannin	GRAS. Not to exceed 2.4 gr/L as SiO_2. Agent use at 30% suspension.

Preservative materials

Material	Used for	Limitations
Ascorbic acid & its isomer Erythorbic acid	Anti-oxidant agent	GRAS.
Dimethyl dicarbonate	Antimicrobial agent	Not to exceed 200 mg/L.
Fumaric acid	Inhibit ML fermentation	Not to exceed 2.4 gr/L.
Sorbic acid & its Potassium salt	To inhibit fermentation	Not to exceed 300 mg/L as sorbic acid.
Sulfur-dioxide & other sulfite agents (sodium and potassium metabisulfite)	(1) Anti-oxidant agent (2) Anti-microbial agent	Not to exceed 350 mg/L.

Acidity correction materials

Material	Used for	Limitations
Calcium carbonate	To reduce excess acidity	GRAS. Acidity shall not be reduced below 5.0 g/L.
Calcium sulfate (gypsum)	To lower the pH	Sulfate shall not exceed 2.0 gr/L expressed as potassium sulfate.
Tartaric acid	To increase acidity	GRAS.
Fumaric acid	To increase acidity	Not exceed 2.4 gr/L.
Lactic acid	To increase acidity	GRAS.
Malic acid	To increase acidity	GRAS.
Potassium carbonate & Potassium bicarbonate	To reduce excess acidity	GRAS. Acidity shall not be reduced below 5.0 gr/L.

Yeast nutrient materials

Material	Used for	Limitations
Diammonium phosphate (and mono)	Phosphate and nitrogen yeast nutrients	GRAS. Not exceed 960 mg/L.
Thiamine	Vitamin yeast nutrient	GRAS. Not exceed 0.6 mg/L.
Yeast extract	Autolyzed yeast nutrient	GRAS. Not exceed 360 mg/L.

The following materials permitted for use as additives in the food industry, were *delisted* from the wine treatment materials: Benzoic acid and its salts, hydrogen peroxide, propyleneglycol, polyvinylpyrrolidone (PVP), and urea.

D. Phenolic Compounds

Phenolic compounds are considered to be one of the most important attributes to wine. They are responsible for wine's color, astringency, bitterness, and partially its taste. They also play a most important part in wine aging. Their chemical reactions in wine are very sensitive to many factors, and consequently they go through continuous chemical changes. The basic information about phenolics is presented in this section, but keep in mind that many observable phenomena in wine, related to phenolic compounds, are not yet clearly understood.

1. Wine Phenolics

This group of compounds is called 'phenolics' because they all contain the phenol molecule in their structure. However, there is no connection between phenol molecule properties and the 'phenolics' characteristics. Phenolics are among the most important chemical groups in plant chemistry in general. We focus here only on the wine aspects of these compounds.

Phenol is the simplest aromatic alcohol, where its ring may be substituted by various functional groups at the different carbon positions (2 to 6) to form many isomers:

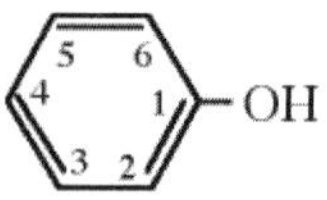

phenol

The major phenolics related to wine and winemaking can be presented in three groups, based on their structural similarity:

- $C_1 - C_6$: the *p-hydroxybenzoic* group
- $C_6 - C_3$: the *cinnamic* group
- $C_6 - C_3 - C_6$: the *flavonoid* group.

The symbol $C_1 - C_6$ denotes a molecular skeleton of a phenolic ring (6-aromatic carbons) + 1 carbon atom. The same meaning and notation relates to the other two groups: $C_6 - C_3$ represents an aromatic ring + 3 carbon chain, and $C_6 - C_3 - C_6$ stands for two aromatic rings connected by 3 carbon chain.

The $C_1 - C_6$ and $C_6 - C_3$ groups are called <u>nonflavonoid</u> phenols, to distinguish them from the third one $C_6 - C_3 - C_6$ the *flavonoid* phenolics. These two classes of phenolics (nonflavonoid and flavonoid) are the major phenolics in wine and also in the fruit of many other plants as well.

- $C_1 - C_6$

This group has the *p-hydroxybenzoic acid* skeleton, and the major derivatives are:

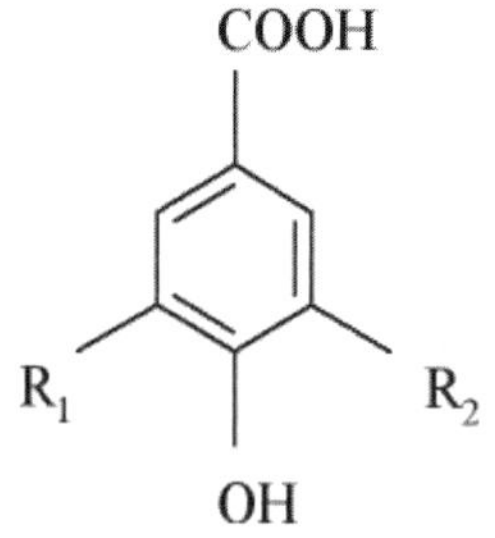

216

Compound	R_1	R_2
p-hydroxybenzoic acid	H	H
p- pyrocatechuic acid	H	OH
gallic acid	OH	OH
vanillic acid	H	OCH_3
syringic acid	OCH_3	OCH_3

- $C_6 - C_3$

This group has the *cinnamic acid* skeleton. Its major derivatives related to wine are:

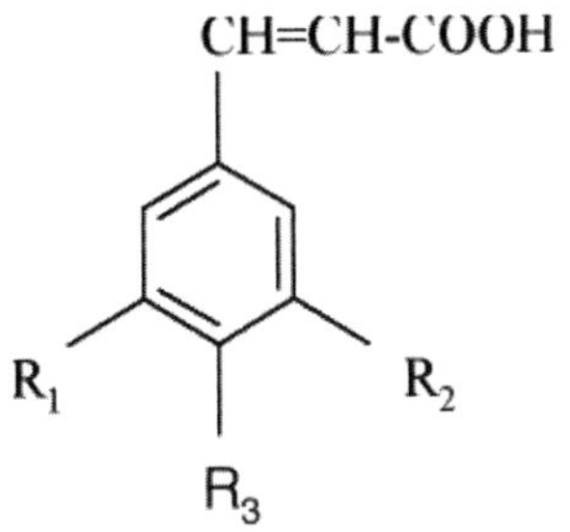

$$CH=CH-COOH$$

Compound	R_1	R_2	R_3
cinnamic acid	H	H	H
p-coumaric acid	H	H	OH
caffeic acid	H	OH	OH
ferulic acid	H	OCH_3	OH
synaptic acid	OCH_3	OCH_3	OH

Both groups of hydroxybenzoic and cinnamic acid derivatives are very rarely found in their acidic form (as they are presented above). Instead, they are usually connected through an esteric bond to alcohols or sugar molecules. For example three major such compounds that are found in white wine are given here:

217

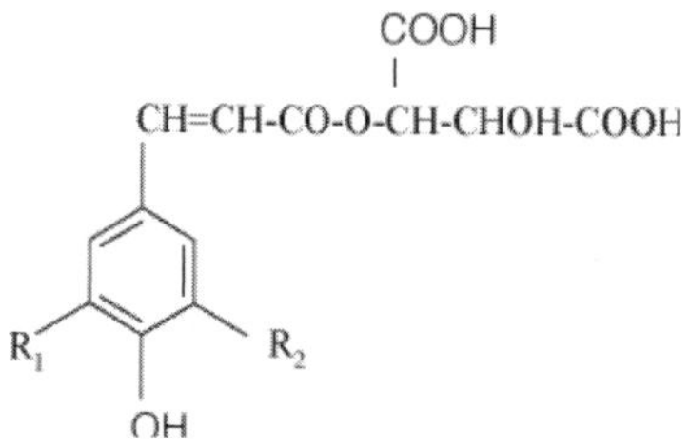

$$COOH$$
$$|$$
$$CH=CH-CO-O-CH-CHOH-COOH$$

R$_1$ R$_2$

OH

cinnamic acid derivative + tartaric acid

Cinnamic derivate	R$_1$	R$_2$	Compound
p-coumaric acid	H	H	coutaric acid
caffeic acid	OH	H	caftaric acid
ferulic acid	H	OCH$_3$	fertaric acid

Another example is gallic acid, which can undergo either an esteric reaction with itself to form *digallic* acid, or a direct C-C bonding between carbons 2 and 6 to form *ellagic acid:*

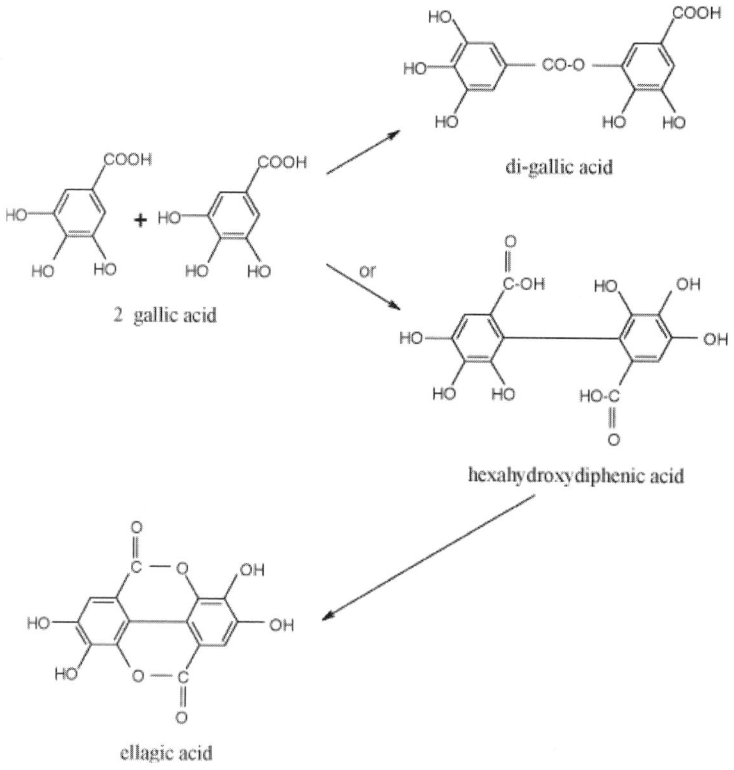

di-gallic acid

2 gallic acid

or

hexahydroxydiphenic acid

ellagic acid

These acids (*gallic* and *ellagic*) are most important in their polymeric form as *oak barrel* extracts, which will be discussed later in this section. In wine production they are extracted from oak barrels to take part in the tannin and flavor in red (and sometimes white) wines.

Free-run juice of all grapes, white and red, contains almost entirely the *nonflavonoid* phenols in about 100–300 mg/L, calculated as gallic acid (GAE, namely, Gallic Acid Equivalent - molecular weight = 170.1)[9].

- $C_6 - C_3 - C_6$

This is the third phenolic group, *flavonoid*, which includes the whole class of plant phenolics, with the $C_6 - C_3 - C_6$ skeleton.

The flavonoids are structured from two phenolic rings (A and B), connected by three-carbon chain, which in most cases is closed by oxygen, forming a heterocyclic ring. In some cases the three-carbon chain is open (a subgroup named chalcones). The characteristics of the different flavonoid groups, are derived mainly by the center heterocyclic ring which classifies the various flavonoids as follow:

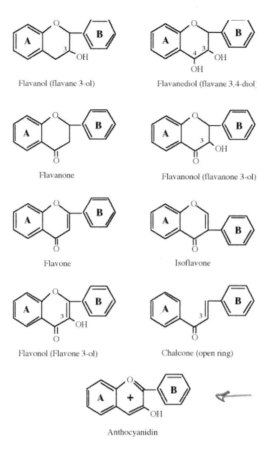

Flavanol (flavane 3-ol)

Flavanediol (flavane 3,4-diol)

Flavanone

Flavanonol (flavanone 3-ol)

Flavone

Isoflavone

Flavonol (Flavone 3-ol)

Chalcone (open ring)

Anthocyanidin

(Note the name changing from *flavane* —> *flavanone* when carbon-4 become a carbonyl, and also *flavanone* —> *flavone* when there is a double-bond between carbons 2-3).

The differences between the many flavonoid compounds that were found in plants and in wines, are based on the center ring structure and on the type of substitutions on the B ring.

Just to mention two of most importance, the catechin and gallocatechin:

catechin

gallocatechin

One of the flavonoid forms (see above) is positively charged and is the basis for a large phenolic subgroup called *anthocyanidin*. Among these compounds are the colored agents in wine, and other fruits as well.

Compound	R_1	R_2
pelargonidin	H	H
cyanidin	OH	H
delphinidin	OH	OH
peonidin	OCH_3	H
petunidin	OCH	OH
malvidin	OCH_3	OCH_3

The compounds listed above are the main flavonoids in grapes, and are responsible for the red color in wine.

The phenolic compounds (*non-flavonoids* and *flavonoids*) in grapes are almost entirely located in the skins and the seeds (although the juice contains some small quantity of it). In white wine production, some of it is extracted from the skin during crushing or through limited skin contact.

In red wine, these compounds are much more extracted from the skin, by maceration during must fermentation. The total phenolics in red finished wine, is in the range of 1000–2000 mg/L as GAE (gallic acid equivalent). Most of them are flavonoids. Rosé wines may contain total phenolics at about 400–800 mg/L, of which 40%–60% are flavonoids. White wine contains about 100–400 mg/L of total phenolics, mostly nonflavonoids, but also some amount of flavonoids (20–40 mg/L), especially catechin. Oak phenolics, which are nonflavonoids (mostly gallic and ellagic acids derivatives) are contained in red wines (by extraction due to barrel aging) in a range of about traces up to 200 mg/L.

Glycosides: Like the hydroxybenzoic and cinnamic acid groups, the flavonoids are seldom not attached to other molecules, and most frequently they are, mainly to sugars. Their *glycosidic bond* is of the form C-O-C between carbons carrying hydroxyls in both the flavonoid system and the sugar molecules. The preferred position in the flavonoid molecule is through carbon-3 (in center ring), and also carbon-5 and carbon-7 in the A-ring. D-Glucose is the most favored sugar molecule, but glycosides with L-rhamnose, L-arabinose and D-galactose are also common. From the sugar point of view, the link is carried through its carbon-1, with a β-configuration (OH above the sugar surface) as the preferred orientation in the sugar molecule.

The glycosidic form of flavonoids gives them greater solubility in water, and enhance their stability against enzymatic oxidation. Free flavonoids (non-glycosidic), are also called *aglycones*. Acidic hydrolysis of phenolic glycosides releases the sugars and the relevant aglycone.

Some examples of different kinds of glycosides are given here:

Mono glucoside:

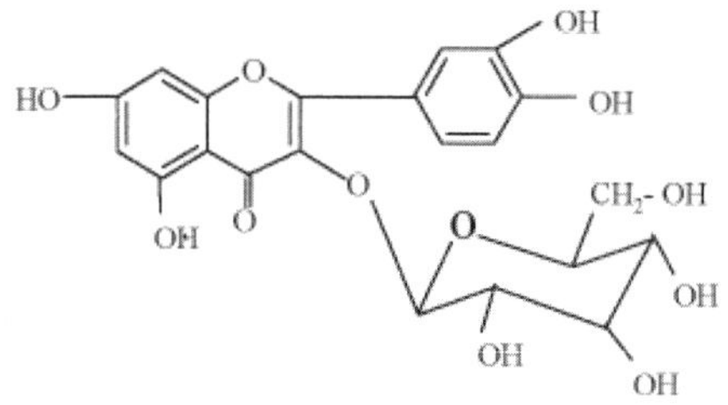

quercetin-3-glucoside

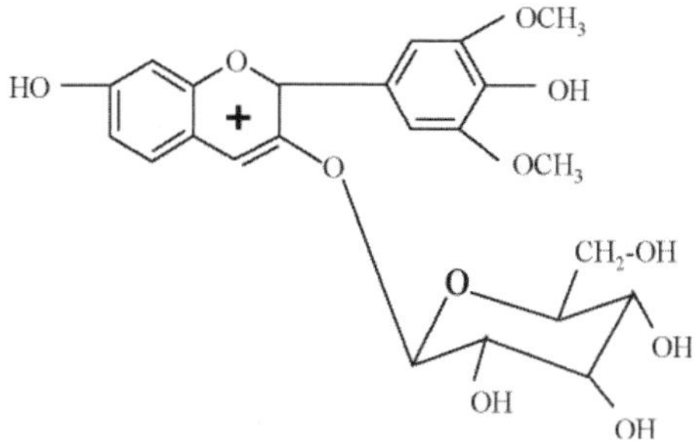

malvidin –3-glucoside

Di-glucoside (the glycoside bonds here are at carbon-3 in the center ring and carbon-7 in A ring):

kempferol-3,7-glucoside rhamnoside

In the last examples note that when anthocyanidin aglycones are glycosided, they are called *anthocyanins* (vs. *anthocyanidin* when they are *not bound* to sugar). Also it is important to mention that *Vitis vinifera* varieties *do not* contain diglycoside anthocyanins, whereas *Vitis labrusca* and its hybrids do. This fact is used to distinguish between hybrids and pure *Vitis vinifera* vines in case of doubt. The flavanes (Flavanol-3 and flavane-3,4-diol) are exceptional in not being glycosidic, but instead they polymerise with another flavonoid molecule to form tannins. More details on this matter will be discussed in the following section.

2. Tannins

The origin of the term *tannin* comes from the leather industry, where plant extractions containing polyphenolic substances were used to produce leather by their reaction with hide.

In solution the tannins can interact with proteins and mutually precipitate. The linkage occurs between the phenolic hydroxyl groups of the tannin's polymer, and the oxygen of the keto-imide (-CO-NH-) in the protein chain. In leather production the hydrogen bond cross-linkage between the polyphenolic hydroxyl and the hide's collagen, creates stability and resistance to water, heat and microbial spoilage. The product of such interaction is commercial leather. In order for a tannin-protein interaction to occur, it needs the suitable structure of high molecular weight tannin. The optimum size range is about 1000–3000 Dalton (a polymer of 3–10 monomer units). In this range of polymer size, it is big enough to cross-link with the collagen chain, and small enough to be able to penetrate into the interfibrillar region to reach the active sites. Nowadays this term is used as a general name for polymers of the vegetable phenolic components, which have certain chemical and physical properties, such as water solubility, molecular weight in the 500–3000 Dalton range, and the ability to interact with proteins and polyamides.

Based on their ability to be hydrolyzed, tannins are classified into two categories: *hydrolysable* and *condensed tannins*.

• **Hydrolysable tannins** are copolymers of gallic and/or ellagic acids with sugar (in most cases glucose). These tannins are also called *gallotannins* and *ellagotannins* respectively. The exact structure of the polymers is not known but by partial hydrolysis one can find segments that contain one glucose unit attached to one gallic acid and two digallic acid molecules such as:

1-galloyl-3,6-digalloylglucose

• **Condensed tannins** are polymers of flavonoids which are condensed mainly through C-C bonds. These bonds are not easily hydrolyzed under regular conditions. The polymerization between the monomers takes place through carbons 4, 6 and 8 in the flavonoid system. The most common condensed tannins are polymers of flavanol-3 (catechin group and their epi-isomes), or the less common flavone-diol 3,4 (leucocyanidin group). They're also called *procyanidins*, or *leucocyanidins*, because under hot and acidic conditions they may dissociate, and by oxidation they can be transformed into cyanidins. An examples of such trimmers, and tetrammer are given here:

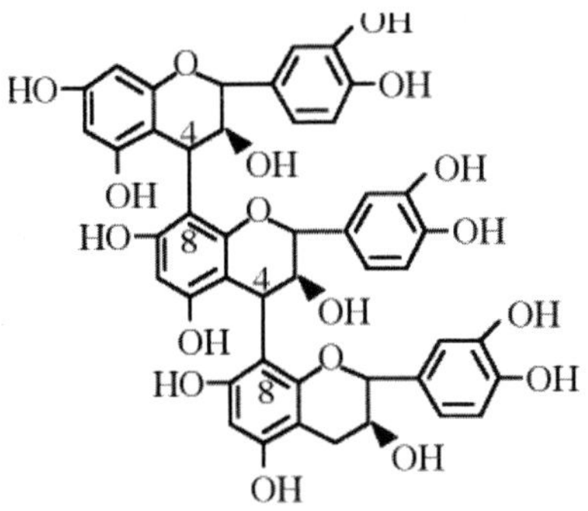

4-8, 4-8 tricatechin

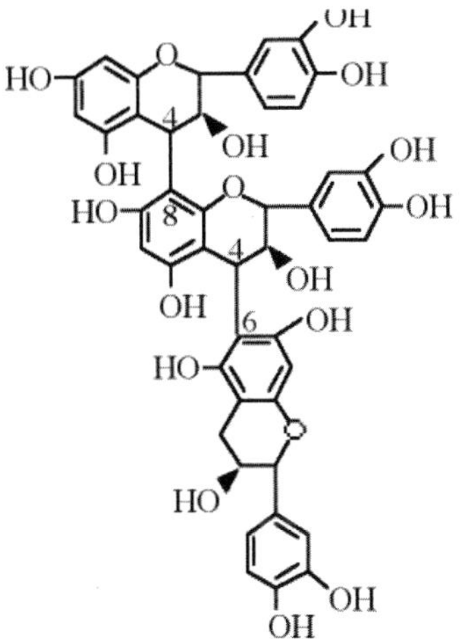

4-8, 4-6 tricatechin

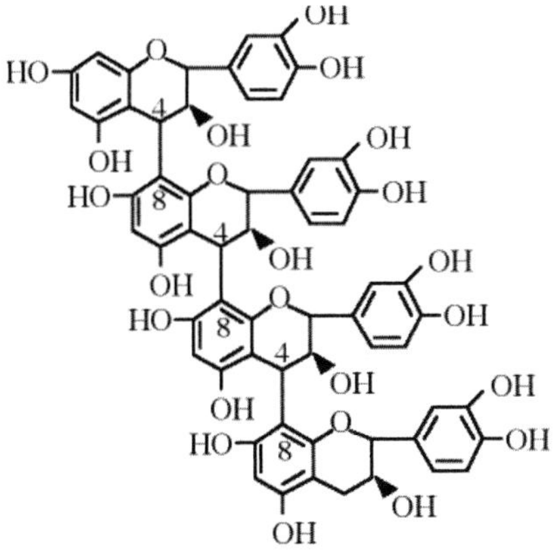

4-8 all tetracatechin

The classification of tannins to *hydrolysable* and *condensed* ones has only historical meaning. Both classes can be hydrolyzed. So the use of these terms is more to differentiate between the main monomers composing the polymers, namely, gallic-ellagic-glucose in the *'hydrolysable tannins'*, and flavanes in the *'condensed tannins'*.

Copolymers composed of a mix of flavonoid and cinnamic acids segments are also known, for example:

gallocatechin gallate

The molecular weight of a monomer is about 300, and the number of units found in wine's tannins is up to about 10. Hence the molecular weight of wine's tannins is between 600–3000. The relative distribution of tannins in wine with respect to their polymerization size changes during wine aging. Also, during aging in oak barrels, hydrolysable tannins are extracted into the wine, a process that has a big impact on the wine's taste and bouquet.

The content of *polymeric phenols* in free-run young white wine is less then 1% of the *total phenolic* concentration, which is about 100–400 mg/L. Upon skin-contact of several days it may increase to 2%–5% of the total in varieties such as Colombard and Sauvignon Blanc, and up to 10%–20% in Semillon and Chardonnay. In red free-run grape juice the polymeric phenols are about 5%–15% (of the total phenolic concentration), and may increase upon long skin-contact to 20%–40%. In aged red wine the polymeric phenols concentration is constantly increasing.

Stems and leaves contain significant concentration of polymeric phenols, up to about 10 gr/kg and 6 gr/kg, respectively. However, during regular wine processing, their contribution to the phenolic concentration upon contact was found to be of small significance.

3. Red Wine Color

Flavylium ion: The compounds, which make red wine look red, are the anthocyanidins. They are well known pigments in plants, responsible for a wide range of colors (almost the full visible spectrum), in flowers, fruits, and leaves. The anthocyanidins contain in their molecular structure the *pyrylium* cation mentioned above:

where R_1 and R_2 are a combination of H, OH and OCH_3.

The actual color in plants depends on various parameters such as the specific anthocyanin, its glycosidic form (anthocyanidin), the concentration in the tissue cells, interaction with metals (aluminum, iron), and on the pH. In most fruits there is one major anthocyanin, which dominates its color. For example: cyanidin (in apple, blackberry, fig, peach, cabbage), delphinidin (in passion fruit, pomegranate, eggplant), cyanidin and peonidin (in cherry, plum, onion). In *Vitis vinifera* grapes all six anthocyanidins mentioned in the previous section are represented: pelargonidin, cyanidin, delphinidin, peonidin, petunidin and malvidin.

pH Dependence of Anthocyanin Color: Anthocyanins in solution are involved in equilibrium with other flavonoids which may have different colors. And because cations are involved, this equilibrium is pH dependent. The following model showing the pH dependency of red wine color was suggested [10]. The anthocyanin cation takes part in two sets of equilibrium:

(1) By a nucleophilic water addition to form *carbinol* which is colorless, and which is also in equilibrium with its colorless to yellow open ring tautomer *chalcone*.

(2) The second equilibrium takes place with the anhydrobase *quinoidal* form, which is violet. The *quinoidal* form is also in equilibrium with its *anionic tautomer* whose color is blue. It is clear then, that the color and intensity in anthocyanin solution will be very much dependent on that particular pigment's equilibrium, which is pH dependent.

1. 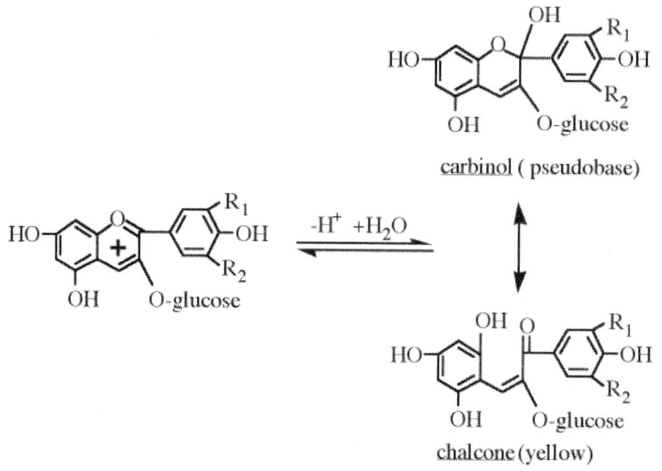
HO
OH
O-glucose
R1
OH
R2
-H+ +H2O
carbinol (pseudobase)
chalcone (yellow)

2. 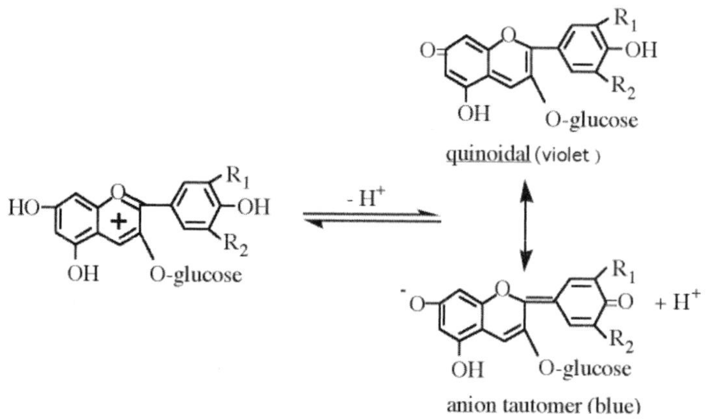
HO
OH
O-glucose
R1
OH
R2
- H+
quinoidal (violet)
anion tautomer (blue)
+ H+

To deal in a more specific case lets look at the major component in red wine pigments, namely, *malvidin glucoside,* whose pK_as for the two equilibriums are known to be 2.6 and 4.25 respectively. Let's summarize the case of malvidin glucoside as follows:

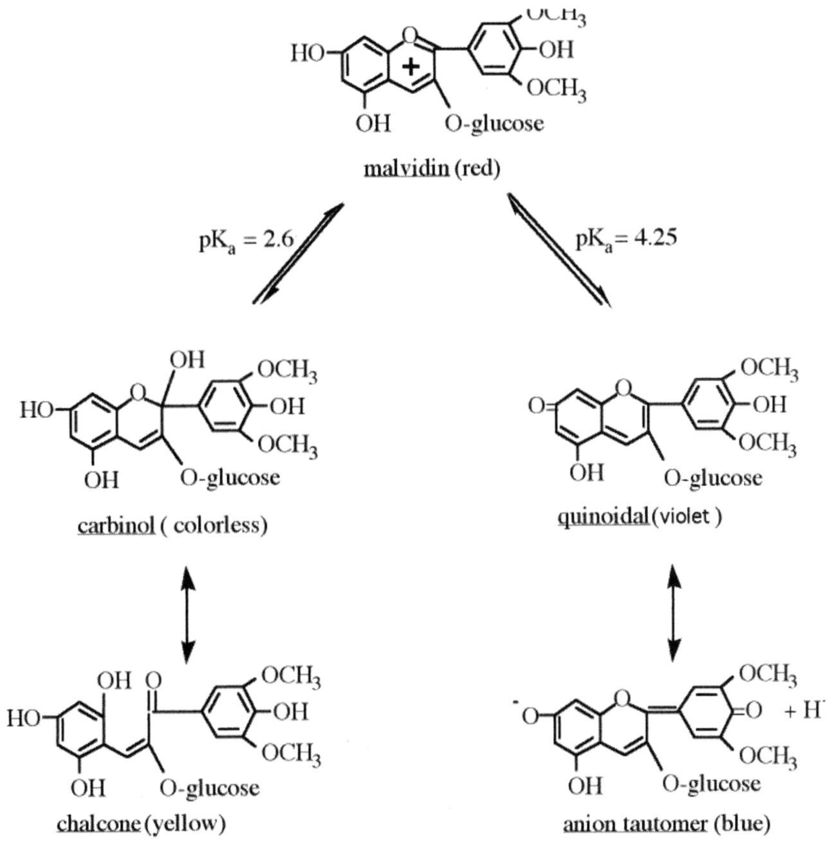

HO
OCH3
OH
OCH3
OH
O-glucose
malvidin (red)
pKa = 2.6
pKa = 4.25
carbinol (colorless)
quinoidal (violet)
chalcone (yellow)
anion tautomer (blue)
+ H

From these equilibrium equations, and the known relation between the various colored parts in the solution, it is possible to calculate the relative concentrations of each component in solution at any pH value, and hence the color and intensity of the solution. The relations are summarized in the equation:

$$\log [F^+]/[F] = pk_a - pH$$

where: $[F^+]$ - is the ionic molar concetration (malvidin in this case).
$[F]$ - is the molar concentration of the other form of the anthocyanin molecule.
pK_a, pH - are the relevant pK_a, and the solution pH.

Based on the above equation, malvidin eqilibrium in low pH range appears as:

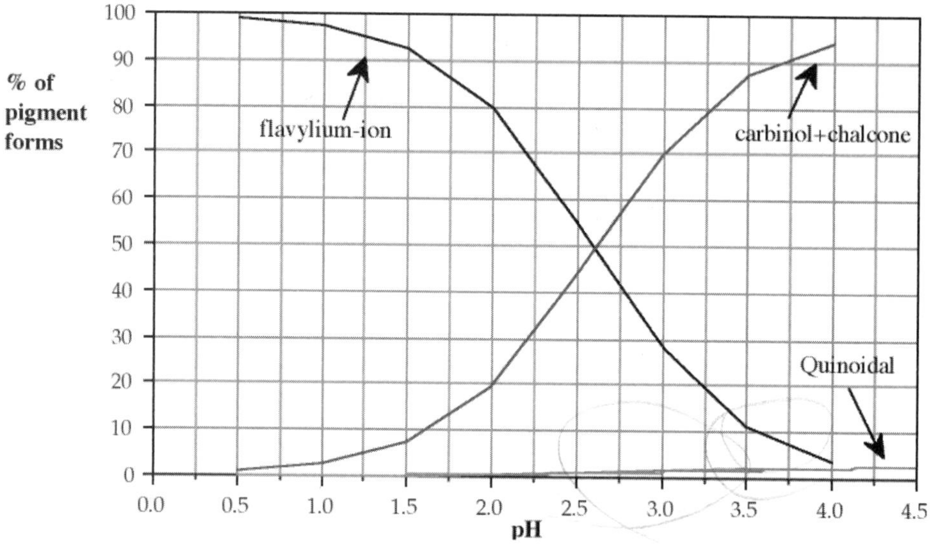

% of pigment forms

pH dependence of malvidin-3 glucoside forms at equilibrium (at 25°C).

Notice how low the red colored form (malvidin-ion) is in the wine pH range (e.g. at pH = 3.25 it is about 20%). On the other hand it is practically 100% at pH below 0.5. This figure shows two significant facts: First, how low is the actual concentration of the red fraction of anthocyanin in red wine, and second, how sensitive is the red color intensity to the wine pH. The lower the pH, the higher is the intensity of its red color.

One remark must be made here with significant viticultural importance. It has been found that in native American vine species, the anthocyanin glucosides are bound to *two* sugar molecules—di-glucoside (see above section-1 in this chapter). The di-glucoside character is genetically transformed to all varieties crossed with the native American ones. A quite simple chromatographic test for mono and di-glucoside malvidin, which is one of the most common anthocyanin in grapes, can verify whether the grapes are pure *Vitis Vinifera* or *Vitis Labrusca* hybrid. (It must be added that this conclusion is not cut and dry, because there are some rare, isolated exceptions). In this connection, it is interesting to compare the equilibrium constants (pK_a) of monoglucoside and diglucoside anthocyanidins. The pK_a values of diglucoside are significantly lower. For example at pH = 2.99 cyanidin 3-glucoside and cyanidin 3,5-diglucoside have pK_as values of 3.01 and 2.19 respectively.

In term of colored forms, their fractions are 51% and 14% respectively at this pH. In another words, it means that at the same concentration of anthocyanin and at the same wine pH, a *Vitis labrusca* red variety wine will be much *less* red-colored than *Vitis vinifera* are.

References

1. Heard, G.M. and Fleet, G.H., "The effect of sulfur-dioxide on yeast growth during natural and inoculated wine," *Aust. & New Zealand Wine Ind. J. 3* (1988) p. 57.
2. Macris, B.J. and Markakis, P., "Transport and toxicity of sulfur-dioxide in Saccharomyces cerevisiae var ellipsoideus," *J. Sci. Food Agr., 25* (1974) p. 21.
3. Yang, H.Y., "Effect of sulfur dioxide on the activity of Schizosaccharomyces pombe," *Am. J. Enol. Vit. 26* (1975) p. 1.
4. a. Rankine, B.C. et al., "Saccharomyces bailii, a resistant yeast causing serious spoilage of bottled table wine," *Am. J. Enol. Vit. 24* (1973) p. 55.
    b. Pitt, J.I., " Resistance of some food spoilage yeasts to preservatives," *Food Tech. In Aust. 26* (1974) p. 238.
5. Schaeffer, A., "Cold temperature sulfiting and stabilization of wines in Alsace," *Vineyard & Winery Management,* May/June (1987) p. 39.
6. a. Beech, F.W. and Thomas, S., "Action antimicrobienne de l'anhydride sulfureux," *Bull. O.I.V. 58* (1985) p. 564.
    b. King, A.D. et al., "Factors affecting death of yeast by sulfur-dioxide," *J. Food Prot. 44* (1981) p. 92.
    c. Heard, G.M. and Fleet, G.H., "The effect of sulfur-dioxide on yeast growth during natural and inoculated wine," *Aust. & New Zealand Wine Ind. J. 3* (1988) p. 57.
7. Vos, P.J.A. and Gray, R.S., "The origin and control of hydrogen-sulfide during fermentation of grape must," *Am. J. Enol. Vit. 30* (1979) p. 187.
8. a. Recht, J.A., "Wine stabilization: chemical and physical stability," *Wine East,* May/June (1993) p. 16.
    b. Recht, J.A., "Wine stabilization: microbiological stability part I," *Wine East,* Jan/Feb (1993) p. 20.
    c. Recht, J.A., "Wine stabilization: microbiological stability part II," *Wine East,* March/Apr 1993) p. 10.
    d. Recht, J.A., "Vinegar," *Wine East,* July/Aug (1994) p. 12.
9. a. Singleton, V.L., "Grape and wine phenolics: background and prospects," from *Grapes and Wine Centennial Sym. Proc.,* in Davis, California, ed by Webb, D.A. (1980) p. 215.
    b. Kantz, K. and Singleton, V.L., "Isolation and determination of polymeric polyphenols in wines using sephadex LH-20," *Am. J. Enol. Vit. 42* (1991) p. 309.
    c. Singleton, V.L., "Tannins and the qualities of wines," in *Plant Polyphenols,* ed. by Hemingway, R.W. and Laks, P.E
10. a. Brouillard, R. and Dubois, J.E., *Am. J. Chem. Soc. 99* (1977) p. 1359.
    b. Brouillard, R. and Delaporte, B., *ibid, 99* (1977) p. 8461.
    c. Brouillard, R. and Chaine, E., *ibid, 102* (1980) p. 5375.

11. Corison, C.A. et al., "Must acetic acid and ethyl acetate as mold and rot indicators in grapes," *Am. J. Enol. Vit. 30* (1979) 130.

12. Margalit, Y., *Concepts in Wine Chemistry,* (1997), chapter III.

Le laboratoire de distillateur liquoriste, d'après Demachy (1775).

APPENDIX

Wine Evaluation

In order to evaluate wine and say anything about it besides "I like it" or "dislike it", a set of tools, concepts and language has to be established. One of the major tools to evaluate sensory response to a stimulus (simple or complicated) is to test it, using people as the measuring instruments. In this section we shall touch very briefly on the basic concepts of wine evaluation. For a wider view of this subject, the reader is referred to the literature.[1, 2, 6, 7]

First, let us define some common terms used in wine evaluation.

Aroma: Is a term used to describe the smell of a wine derived from the grape. Each grape variety contributes a specific aroma (simple or complicated) to wine made from it, which characterizes its varietal uniqueness. Some varietal aromas are very powerful and easily recognized, while others are very weak and unspecific. Not only the grape variety is characterized by the wine aroma. The growing site of the grapes (the 'Terroir') may also contribute its uniqueness to the wine specific aroma.

Bouquet: Is the term for the smell of wine derived from its processing and history. This includes fermentation conditions, cellar operations and aging (barrel and bottle). The bouquet continuously changes during the life of a wine, as it progresses from a young fresh wine to an old aged one. Bouquet is very sensitive to most cellar operations, and to the storage conditions of the wine.

Taste: Is the tongue response to sweet, salt, sour and bitter stimulus. Three of these are found in wine (sugar, acids and bitter compounds) and are partially responsible for its taste perception.

Flavor: This general word is used in wine evaluation language in the narrow meaning as the "aroma sensation of the wine in the mouth", in parallel to its nose perception, called "aroma".

Body: Is a term, which belongs to the "touch" perception, which is also known as "feeling". Touch is perceived by the whole mouth, and is stimulated by factors such as temperature, hot spices, burning due to alcohol, tactile feeling due to dissolved gases, solid texture, and viscosity in liquids. In wine evaluation, two terms are of great importance, namely, *body* and *astringency*. The *body* is defined as the "fullness" or the "roundness" touches of the wine in the mouth. It is a sensation caused by the alcohol content, sugar content, and other dry extract compounds in the wine.

Astringency: Is the "squeezing" feeling in the mouth due to the presence of tannins. It is caused by the reaction between tannins and the proteins in the mouth, which creates a "roughness" sensation in the mouth tissue.

There are two main aspects in evaluating wines. One is called *descriptive analysis*, and the other *rating tests*. The two aspects are presented in this section.

1. Descriptive Analysis

The object of this analysis is to create a common language and terms for describing the characteristics and quality of a beverage. The initiative to develop such common flavor vocabulary started in other food fields, while the major effort in alcoholic beverages was made by the brewing industry back in 1974-5. Terms had been defined to describe the possible flavors and sensations of beer. The terms were organized in a two-tier wheel, where the first is very general such as 'mouth feel', 'aromatic', 'nutty' 'musty' etc. The terms in the second tier were more specific, for example 'moldy,' 'leathery,' and 'papery' for the 'musty' term in the first tier. The concept of an "aroma wheel" of beer flavor has become a useful tool in beer research.[3] A few years later, in 1979 the concept was then expanded to analyze whisky as well.[4,5]

In order to do descriptive analysis of wine, the various terms for describing wine must first be defined. These terms have to represent, as fairly as possible, the wine's domain of smells, flavors and faults. Also, they have to be known and familiar to the taster's senses. Second, these terms have to be organized into groups, according to the associations of the members in that group or category. Then, the terms can serve as the vocabulary in the wine evaluation language. Such terms were defined, and standard solutions of components, which can be used for training people, were developed[6,7] in 1984–87. The results were published as what is called the "Wine Aroma Wheel," which contains in its inner cycle the very basic terms that may associate with the wine under study, such as: *fruity, vegetative, nutty, woody, earthy, floral*, and others. The second cycle separates some of these terms into subgroups, for example, the *vegetative* is sub grouped into *fresh, canned*, and *dry*. The outer cycle is much more specific in describing the associative smell, e.g. *grass, bell pepper, mint, asparagus, green olive, hay, tobacco*, and others. On smelling or tasting wine, in many cases it is possible to associate the major impression of the wine with some of the terms in the wheel, and maybe with some minor associations as well. The use of these terms helps to communicate with other members of the tasting panel. The "Wine Aroma Terminology Wheel" is shown in the following figure[6, 7]:

Modification of a Standardized System of Wine Aroma Terminology

A.C. NOBLE, R.A. ARNOLD, J. BUECHSINSTEIN,
E.J. LEACH, J.O. SCHMIDT and P.M. STERN

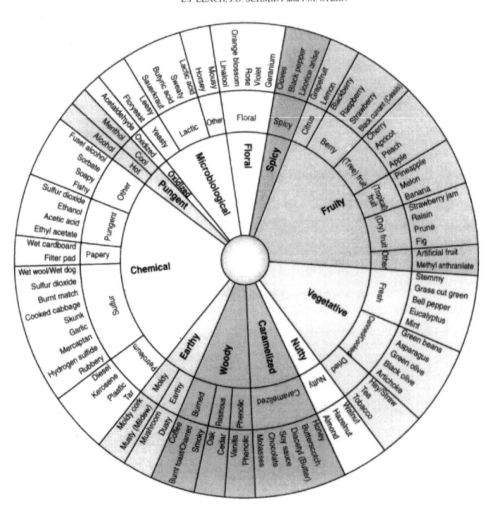

2. Rating Tests

In a *rating test* the question asked is "on a given scale where would you put each of the tested samples?" Before such a question can be asked, a set of basic rules has to be established. The first one is, what is the scale? It can be a numerical scale (e.g. from 1 to 10), or a graphic line at a given length (e.g. 10 cm), which is marked 'low' at left side and 'high' at right. The judge is asked to assign a number or equivalent length to his/her perception on the *intensity* of a certain attribute under study. The panel collective numbers would produce a mean, which represents the panel's judgment. In using such scaling techniques, some problems may arise. First, the individual perception on any issue can be widely ranged between the judges at any time, or specifically at the time of the test.

The second difficulty is that people express differently their perceptions on a scale. For example many people do not use the scale edges, avoiding extreme positions. By doing so they narrow the actual range of the rating. These people do not use the scale in a linear way. Values in the middle are more widely used, so the middle values are more condensed, which makes it look as if the scale is not linear. On the other hand, some people always take extreme attitudes, avoiding the use of the middle scale. In general, using people as measuring instruments is really difficult, unless at least they can be 'calibrated' (as is done in any case with instruments). To partially overcome these difficulties, reference samples can be given at the tasting, and in many cases they are anchored to certain positions on the scale. For example in a very simple sour test (to compare the sourness of a list of samples), the very low and the very high concentrations of sour samples are presented, and are referred to as "1" and "10" on the scale. By doing so, two anchors are set on the scale. One anchor reference is also possible, for example in the middle of the scale (numeric or graphic). Using anchor references helps to more homogenize the panel responses on a common basis, and to set a more reliable scaling spread.

The size of the scale is also important. A scale field of three (low, medium and high, or bad, fair, and good) is very easy and convenient to use, but on the other hand it is not sensitive enough to express higher resolution of the sensations that people feel and are able to express. A field of ten could be much better, but for many people, as said before, it is too wide. Experience shows that a field of 5 to 7 rating values in the scaling rate would be the most suitable, assuming the judges would really use the full scale.

Rating test for evaluating one sensory dimension is relatively simple after training the panel to do the specific task of the rating. For example, rating the relative sourness, sweetness, bitterness, saltiness, astringency, etc. of solution. All these sensations are compared by their intensity. The results also can be absolutely verified by their concentration in the solution. But how can we evaluate a sample like wine, which is completely different by two aspects: One, there are many parameters, to evaluate within the same sample. These parameters are mixed together and interact with one another. The perception of a mixture of parameters is much more complicated than that of individual ones. Secondly, and very importantly, in evaluating wine, most of the parameters are not quantitative ones, but qualitative. The question here is "how *good* is the smell of the sample" instead for example, "how sweet it is." Also, the taster has to find out the unique characters in the sample, instead of relating it to a given parameter. The evaluation has to rate the wines as a "whole" and to compare the wine samples by rating them in the order of "how good" they are. In addition, the sample should not be rated only within the narrow boundaries of the given list of tested samples, but should also be related to the whole domain of similar wines in the world. All these points are brought here, (in a very short notes), in order to understand that evaluating wine is definitely not simple, and not so accurate as might be impressed.

Also, we did not mention here the standard conditions of the tasting, which can completely alter the results (e.g. glasses, temperature, light, blind tasting procedure, number of wine samples at one tasting session, score tasting procedure, result evaluation procedure, and more).

Facing so many difficulties, some attempts were made to facilitate wine evaluation, which we shall mention here. We shall not deal with technical procedures (which are most important), but rather with the heart of the question, namely, the rating score. The principle in the rating score is to break down the major organoleptic responses (parameters) and evaluate them separately for their contribution to the wine under study, by giving them a numeric score. Each parameter could have a maximum score according to its assumed importance (weight) to wine quality. By summing the individual scores, the total wine quality score can be evaluated. There were many score tables offered in the literature. One of the most recognized is the *Davis scorecard,* which was developed at the University of California at Davis. The original and the modified Davis scorecards are shown in the following table[2]:

Original Score		Modified Score	
parameter	**weight**	**parameter**	**weight**
appearance	2	appearance	2
color	2	color	2
aroma, bouquet	4	aroma, bouquet	6
volatile acidity	2	astringency	1
total acidity	2	total acidity	2
sweetness	1	sweetness	1
body	1	body	1
flavor	2	flavor	2
bitterness	2	bitterness	1
general quality	2	general quality	2
Total	20	**Total**	20

Rating: 20–17 superior
16–13 standard
2–9 below standard
8–1 unacceptable

As the wine industry rapidly changed in California during 1950–1970, the original card was changed to the modified scorecard, which reflected the improvements in wine quality at that period. Although the Davis scorecard was widely used for many years, it has some serious weak points, which limited its full use:

(1) Too high a weight was given to the appearance of the wine (20% of the total quality value).

(2) Too low a weight was given to the most important parameters, namely, aroma, bouquet and flavor (only 30% in the original table, and 40% in the modified).

(3) How to relate the term "sweetness" is not clear. Is it a 'good' parameter, which contributes to wine quality or not ?

(4) Absence of "volatile acidity" in the original card gave a wine 10% quality score, which looks odd in modern winemaking. It was then omitted in the modified card.

(5) The same argument for "bitterness" which is not justified in today's technology, and was not omitted from the modified card.

(6) The term "astringency " like sweetness does not specify if it is 'good' or 'bad,' and how to relate it.

(7) All of these insignificant parameters, or better to say their absence in wine as indication of quality (sweetness, astringency, bitterness), which sum up to 15% in the modified card, are credit given to the wine quality just because of "good behavior."

(8) Too low a weight for "general quality" which include the overall impression of the wine as an integral unity such as the balance, aging, after taste, off-flavor or other 'pro's and 'con's.

(9) Also most important is the range of the rating that each parameter is scaled (between 1 or 2 points, or at most 1/2 point), which does not leave enough space to express the full potential differences in each parameter. (Of course other fractions can be used, e.g. 1/3, 3/7 etc. but then many people will have some difficulties in summarizing them).

The use of the Davis scorecard was re-evaluated after 14 years of use by expert panel members. [8]

It took several years of experience to adjust and to study the 20-point scorecard. In practice this 20-point card showed that in most cases the numbers on the card span between 12 and 16, as if the whole scale was shrunk into this range. It does not make wine evaluation easy if the rating falls into such a narrow range. Other attempts were also made to develop wine scoring cards with different approach to the problems of this matter.[2, 9]

Facing the difficulties in forming any wine tasting card, we challenged the Davis card by proposing a new card, which will include the following principles:

(1) The appearance (color and clarity) would have maximum of 15% of the total score.

(2) Aroma, bouquet and flavor will have 45% of the total score.

(3) A new parameter named "defect" is included which will contain any major and minor defect that exists in the wine. It includes for example those 'insignificant' parameters mentioned above in Davis card critics. And any attribute which reduces wine quality such as moldiness, corkiness, oxidation, too oaky when not supposed to be, astringency when it is too high, bad after taste, mousy, and any defect which can be found on tasting. The evaluation of this "defect parameter" is different than the others and will be discussed later.

(4) A "general quality" parameter (weight 20%), which will allow the judge to express his/her personal opinion on the wine quality, such as the wine's balance, mouth feel (body), aging, and after taste.

(5) The *general quality* and *defects* parameters together weights 30% of the total. It gives the judge a wide range of freedom to appreciate a good wine or to depreciate it, when it gets too many quality points on less important parameters (which happen in many cases in simple well-made wines).

(6) The range of scoring *each* parameter was expanded to span from 0 ——> 5. The taster deals only in a field of six (from 0 to 5), which gives enough resolution without going into 1/2 or 1/3 of a point, and it is not too big to get lost in a too wide range. Each parameter has its own *weight value* according to the principles listed above, namely, to its relative importance in the wine quality rating. Therefore, the rating of each parameter is multiplied by its weight, and the total is summarized.

The total parameters *weights* is 20, and as the maximum rating of each individual parameter can be 5, the maximum wine score can therefore be 20 x 5 = 100. The card was presented in ref. 1 and is shown here in the following table:

	No.				
Date _____ Name _____	Wine Winery Vintage Price				
Appearance x2 turbid, clear, brilliant					
Color x1 light, normal, dark					
Nose x6 varietal aroma & bouquet					
Flavor x3					
Acidity x2 low, balanced, too high					
Defects x2 acetic, oxidized, moldy, too oaky, sulfide, bitter, corky, bad after taste					
General Quality x4 body, astringency, age, balance, after-taste					
Total					

Score each parameter 0 ——▶ 5 then multiply by weight factor

Rating: 100–96 superior
                  96–86 excellent
                  85–71 good
                  70–56 standard
                  55–41 below standard
                  40–0 unacceptable

To use the card, each parameter is rated between 0–>5 and multiplied by its weight factor. The *defects* parameter is rated a full 5 points if *no defect* is found in the wine, and if one or some defects are present, it is degraded in rating according to the seriousness of the defects, down to zero rating. The use of the *general quality* parameter is to express the *general* and *integral* impression of the wine, and emphasize any special issue the taster wants to address (for good or for bad) by the appropriate score. After rating each parameter, the columns sum enable to compare the wines score. For blind tasting, the wine's number is noted on the top of each wine column. The details of the wines (winery, vintage, price), can be disclosed after tasting.

After a few years of use, we found this card was not fulfilling some of its expectations:

(1) The look of the wine has very much less importance then it has been given in the card (15% of the score). The brilliancy of white wine has nothing to do with its quality, unless it is a real defect that had occurred in the bottle. This defect will show up anyway in its smell and in the mouth and will find its evaluation there. The technology nowadays in producing clear white wine is very common, and it has nothing to do with wine quality. In red wines the clarity is difficult to see because of the red color, mainly in dark heavy wines. So, it is irrelevant to the wine quality. The hue of the wine (white and red) reflects its age or its oxidation state. It is not a matter of quality. If the wine is over-oxidized, it will be very clearly shown in its bouquet and taste. If the wine is red-brown or white-yellow because of age, it will again be felt by the nose and mouth senses. In both cases the wine looks the same and no real conclusion can be deducted from its look. In red wine production, it is

quite common now, not to filter it before bottling. How do we relate to it in the scoring card? And sediments in aged red wine bottle, are normal and do not show any connection to its quality. So, we find these parameters of appearance and color, irrelevant, and should be eliminated.

(2) The acidity parameter is really not a quality term by itself. In a one-dimensional test it can only be judged by its concentration or intensity. Its real quality contribution to wine is by its balance with other parameters, such as sweetness, astringency and body. So, there is no justification to take it as quality parameter, and score wine just for its acidity, but instead, to relate to it as part of the balance in the tested wine.

(3) As a matter of fact, we do not think that one should relate to *any* specific attribute such as acidity, bitterness, sweetness, astringency, body, oakyness, etc, as quality parameters. Because neither of these attributes, stand by itself as quality parameters. Only their combination in the evaluated wine determines its quality. Therefore, instead, we suggest in evaluating wine quality, to focus on the *harmony* and *pleasant sensation* we feel by our nose and by mouth. The various parameters can be judged by their contribution to this harmony, for good or for bad. By doing so, we retreat back from the old method of differentiating wine quality to its individual factors, and instead, evaluate wine quality as an integral sensation stimulator.

(4) Also, there is not any relation to the specific *characteristics* (descriptive) to the tested wines in any scorecard. This part of evaluation is of most importance, and should be included in the scorecard.

Based on these arguments, we suggest a new scorecard containing four parameters. Three of them are positively perceived, and the fourth is a negative one, because unfortunately, negative attributes exist in wine. The parameters are: *Nose, Mouth feel, Harmony,* and *Negative Attributes.* Each of them contains of course the familiar factors, which compose wine quality.

Nose: The smell sensation, which includes in it the varietal aroma and bouquet. Its weight is x7.

Mouth feel: The 'taste' sensation in the mouth, which contain the varietal flavor, bouquet, body, and acidity. Its weight is also x7, equal to the *Nose*.

Harmony: The integral impression of the evaluated wine, which include its appearance, balance, complexity, aging, after-taste and uniqueness. Its weight is x6.

Negative Attributes: The negative parameters, which include the factors that reduce wine quality, such as volatile acidity, oxidation, bitterness, over-oak, repulsive off-flavor, corkyness, opaque color etc. This parameter weight is x7. It has heavy weight in order to emphasize the heavy negative impact that exists in wine quality due to defects, in spite of the other positive parameters. The sum of the weights of the three positive parameters is 20. Each one of them can be scored between zero to five, hence the maximum score might be 20x5 = 100 points. The negative parameter score range is from zero up to -7x5 = -35 points, which are *deducted* from the total score.

In this way, the ultimate wine with no defects could be rated 100 by summing up its three quality parameters. Defects are not counted unless we notice them, and then they reduce the score as needed to bring the wine quality evaluation to its actual rating. No 'good behavior' exists in the rating, but rather 'bad behavior' is punished if found. We feel that using these scoring principles to rate wines will better express the quality and technology of the current wine industry.

The scoring chart also contains a column for describing the wine by using terms from the Wine Aroma Wheel. We believe this scorecard is simple and clear to use. It emphasizes the main parameters of wine quality and its characteristic descriptions, and will be helpful in the field of wine evaluation. The scorecard is shown in the following figure. Each card is for one wine sample.

WINE TASTING EVALUATION

Date _____

Variety _____   Vintage _____   Wine No. _____

	Weight	Characteristics	Score (1 → 5)	Value Weight x Score
Nose Varietal aroma Bouquet	x7			
Mouth feel Flavor (varietal & bouquet), Acidity, Body	x7			
Harmony Color Balance Complexity Aging After taste Uniqueness	x6			
Negative Attributes Color, Acetic, Oxidized, Moldy, Bitterness, Corky, Repulsive off-flavor, Oaky	-x7			

Total _____

REFERENCES

1. Margalit Y., *Concepts in Wine Chemistry,* Chapter IV. (1997).

2. Amerine, M.A. and Roessler, E.B., *Wines, Their Sensory Evaluation,* (1976).

3. a. Clapperton, J.F. et al., "Progress towards an international system of beer flavor terminology," *Master Brewing Assoc. of Am. Tech. Quat. 12* (1975) p. 273.
   b. Meilgaard, D.S. et al., "Reference standards for beer flavor terminology system," *J. Am. Soc. Brew. Chem. Proc. 40* (1982) p. 119.

4. Shortreed, G.W. et al., "The flavor terminology of Scotch whisky," *Brewer Guardian 108 (11)* (1979) p. 55.

5. Piggott, J.R. and Jardine, S.P., "Descriptive sensory analysis of whisky flavor," *J. Inst. Brew. 85* March (1979) p. 82.

6. Noble, A.C. et al., "Progress towards a standardizing system of wine aroma terminology," *Am. J. Enol. Vit. 35* (1984) p. 107.

7. Noble, A.C. et al., "Modification of a standardized system of wine aroma terminology," *Am. J. Enol. Vit. 38* (1987) p. 143.

8. Ough, C.S. and Winton, W.A., "An evaluation of the Davis wine score card and individual expert panel members", *Am. J. Enol. Vit. 27* (1976) p. 136.

9. Rankine, B., "Roseworthy develops new wine score card," *The Australian Grapegrower & Winemaker,* Feb (1986) p. 16.

Bibliography

1. *Knowing and Making Wine,* E. Peynaud, J. wiley, New York, 1984.
2. *Principles and Practices of Winemaking,* R. Boulton *et al,* 1995.
3. *Concepts in Wine Chemistry,* Y. Margalit, The Wine Appreciation Guild, San Francisco, 1997.
4. *Chemistry of Flavor,* A. Waterhouse & S. Ebeler, 1999.
5. *The Microbiology of Wine and Vinifications (vol. I, II),* P. Ribereau-Gayon, *et al,* 2000.
6. *Production Wine Analysis,* B. Zoecklein, *et al,* Avi Books, New York,1994.
7. *Wine Microbiology and Biotechnology,* G. Fleet, ed., 1993.
8. *Wine Science,* R.S. Jackson, Academic Press, New York,1994.
9. *Vine, Grapes and Wine,* J. Robinson, Albert Knopf, New York,1986.
10. *Grape Growing,* R. J. Weaver, 1976.
11. *Viticulture (vol. I, II),* B.G. Coombe & P.R. Dry, ed., Wine Titles, Adelaide,1992.
12. *General Viticulture,* A.J. Vinkler, *et al.*
13. *Handbook of Enology (Vol. I, II),* P. Ribereau-Gayon, *et al,* J. Wiley, New York, 2000.
14. *Terroir,* J. Wilson, University of California Press, Bereley,1998.
15. *Sunlight into Wine,* R. Smart & M. Robinson, 1991.
16. *The University Wine Course,* M. Baldy, The Wine Appreciation Guild, San Francisco,1997.
17. *The Taste of Wine,* E. Peynaud, The Wine Appreciation Guild, San Francisco,1987.

Index

Ethyl octanoate, 73
Eugenol, 143
Evaporation, through barrels, 127, 134–36, 139

F

Fatty acids, as fermentation inhibitors, 64, 84
Faults
    chemical, 191–200
    microbial, 186, 201–7
    in wine rating tests, 242, 246, 247
    See also Spoilage
Federal Alcohol Administration Act, 207
Fermentation, 47, 57–79
    bottle fermentation, 78–79, 85–86, 204
    inhibitors, 61–65, 184–86. *See also specific inhibitors*
    management procedures, 59–61
    preventing, 78–79, 186
    safety caution, 61
    simultaneous yeast and malolactic fermentation, 84–85
    See also Malolactic fermentation; Yeast fermentation
Fermentation problems, 60
    bentonite fining and, 103
    stinking fermentation, 60, 62, 193–94
    stuck fermentation, 57, 61–65, 103, 186
Ferric phosphate, 192
Ferric tannate, 192
Ferrocyanide, 192, 213
Fertaric acid, 218
Ferulic acid, 25, 165, 217, 218
Filter pads, 36
Filtration, 91, 113–18
    at bottling, 173–74
    centrifugation, 118
    DE filters, 117–18
    lees filtration, 36–37
    membrane filters, 116–17
    over-filtering, 199
    pad filters, 113–16
    pumping caution, 118
    sterile, 79, 113, 206
Fining, 101–12
    before fermentation, 103
    combining with cold stabilization, 97
    nonspecific, 110–11
    over-fining, 199
    pectolytic enzyme treatments and, 32, 67
    protein stabilization, 101–5, 194
    for tannin reduction, 105–10
Fining agents, 111–12, 213
    activated carbon, 109, 110–11, 112, 213
    bentonite, 97, 102–4, 105, 108, 111, 115, 194, 213
    casein, 108, 112, 213

    egg white, 109, 112, 213
    gelatin, 105, 106–7, 111, 213
    isinglass, 108, 112, 213
    milk, 108, 213
    PVPP, 109–10, 112, 213
    silica gel (kieselsol), 104–5, 107, 111, 213
Flat taste, 39, 71
Flavanes, 223
Flavanols, 25, 220, 223, 224
Flavone-diol 3,4, 224
Flavonoids, 26, 219–23, 224
Flavonol haze, 198
Flavor
    barrel aging and, 141–45. *See also* Oak phenolics
    barrel fermentation and, 139, 145
    bentonite fining and, 103
    blending to correct, 119, 120
    carbon fining and, 110
    defined, 235
    harvest timing and, 13
    malolactic fermentation and, 82
    skin contact and, 25, 26
    See also Aroma; Phenolic compounds; *specific flavor characteristics*
Flocculation capability of yeast, 55
Flor sherry yeasts, 71, 203–4
Formic acid, 66
Free-run juice, 32–33, 35
Free SO$_2$
    vs. bound SO$_2$, 186–88
    excessive, reducing, 200
    pH distribution, 180–83
    remaining after fermentation, 189
French Colombard, 119
French oak, 129, 141, 145–49
Fructose, 73, 74
Fruit set, 3
Fruit wine, legal definition, 208
Fumaric acid, 42, 83, 205, 212, 214
Fungal agents and infections
    fungal growth on corks, 166
    SO$_2$ and, 186, 189
    vineyard chemical prevention, 16
    See also Yeast; *specific fungi*
Fusel oils, 67–69

G

GAE (Gallic Acid Equivalent), 219
D-Galactose, 222
Gallic acid, 25, 140, 221
    chemical structure, 217
    oak astringency and, 141, 147
    polymeric forms, 218–19
Gallocatechin, 220, 226
Gallotannins, 134, 224

filtration, 115
fining, 103, 105, 107, 109, 110
glycerol content, 69
inoculation and fermentation procedures, 59, 63
methanol content, 67
oxidation, 23–24, 26, 90, 124, 189, 199
polymeric phenolic content, 226
pressing, 23, 24, 26, 28, 35
production yields, 15
racking during fermentation, 62
recommended must acidity range, 40
skin contact, 23, 24, 25–28
SO₂ additions, 189
storage temperature, 124
sur-lees aging, 139–40, 145
tank space needed during fermentation, 18, 59
tank storage, 124
tannins in, 105–6
time required for fermentation, 49
total phenolic content, 221
typical pH levels, 9
Wild yeasts. *See* Natural yeasts
Wine evaluation, 235–47
definitions, 235–36
descriptive analysis, 236–37
rating tests, 238–47
alternative scorecards, 242–47
Davis scorecard, 240–42
Wine faults. *See* Faults; Spoilage
Winery maintenance
preparation for harvest, 18–20
sanitation, 20, 121–23, 124
Wine storage, 124

Y

Yeast, 19
alcohol tolerance, 54, 56, 184
characteristics, 54–55
cold tolerance, 59
dosage, 55, 58

filtering out yeast cells, 116
flor sherry yeasts, 71, 203–4
hydrogen sulfide production and, 195
inoculation, 58–59, 64–65
killer yeast, 56–57
SO₂ tolerance, 184–86
species and strains, 54–57, 59, 184. *See also specific strains*
as spoilage cause, 203–4, 205–7
storage temperature, 55
unviable, 64–65
Yeast extract, 62, 64, 215
Yeast fermentation, 57–61
barrel fermentation, 19, 49, 139–40, 145
botrytised wines, 75–76
byproducts, 53, 55, 60, 65–73, 189
carbonic maceration, 31
chemical formulas, 6, 53
heat release during, 47
inhibitors, 16, 46, 47–48, 61–65, 78–79, 184–86
lag period between inoculation and fermentation start, 48, 63, 185
natural yeasts, 55–56, 184, 195
phases in, 58
residual sugar, 53, 73–79
stinking fermentation (hydrogen sulfide formation), 60, 62, 193–94
storage and, 55
stuck fermentation, 57, 61–65, 103, 186
sugar consumption, 53
techniques for stopping fermentation, 76–78
temperature and, 47–49, 59, 62–63, 69
typical times required, 49
Yeast ghost, 64
Yeast nutrients, 62, 64, 215

Z

Zero racking, 90
Zinfandel, 29, 30